Lectures in Mathematics
ETH Zürich
Department of Mathematics
Research Institute of Mathematics

Rolf-Peter Holzapfel

The Ball and Some Hilbert Problems

1995

Birkhäuser Verlag
Basel · Boston · Berlin

Author's address:

Rolf-Peter Holzapfel
Humboldt-Universität zu Berlin
Fachbereich Mathematik
Unter den Linden 6
D-10099 Berlin

A CIP catalogue record for this book is available from the Library of Congress, Washington D.C., USA

Deutsche Bibliothek Cataloging-in-Publication Data
Holzapfel, Rolf-Peter:
The ball and some Hilbert problems / Rolf-Peter Holzapfel. –
Basel ; Boston ; Berlin : Birkhäuser, 1995
(Lectures in mathematics)
ISBN-13:978-3-7643-2835-1 e-ISBN-13:978-3-0348-9051-9
DOI: 10.1007/978-3-0348-9051-9

Softcover reprint of the hardcover 1st edition 1995

Printed on acid-free paper produced of chlorine-free pulp ∞

ISBN-13:978-3-7643-2835-1
ISBN-13:978-0-8176-2835-2

9 8 7 6 5 4 3 2 1

Contents

Preface

As an interesting object of arithmetic, algebraic and analytic geometry the complex ball was born in a paper of the French Mathematician E. PICARD in 1883. In recent developments the ball finds great interest again in the framework of SHIMURA varieties but also in the theory of diophantine equations (asymptotic FERMAT Problem, see ch. VI). At first glance the original ideas and the advanced theories seem to be rather disconnected. With these lectures I try to build a bridge from the analytic origins to the actual research on effective problems of arithmetic algebraic geometry.

The best motivation is HILBERT's far-reaching program consisting of 23 problems (Paris 1900) "... one should succeed in finding and discussing those functions which play the part for any algebraic number field corresponding to that of the exponential function in the field of rational numbers and of the elliptic modular functions in the imaginary quadratic number field". This message can be found in the 12-th problem "Extension of KRONECKER's Theorem on Abelian Fields to Any Algebraic Realm of Rationality" standing in the middle of HILBERTS's program. It is dedicated to the construction of number fields by means of special value of transcendental functions of several variables. The close connection with three other HILBERT problems will be explained together with corresponding advanced theories, which are necessary to find special effective solutions, namely:

7. Irrationality and Transcendence of Certain Numbers;
21. Proof of the Existence of Linear Differential Equations having a Prescribed Monodromy Group;
22. Uniformization of Analytic Relations by Means of Automorphic Functions.

For the convenience of the reader the full translated text of these four HILBERT problems is presented in Appendix 2.

Acknowledgement

I would like to express my gratitude to J. MOSER and G. WUESTHOLZ for their invitation to hold the lectures at the ETH Zürich. I also thank H. SHIGA, J. ESTRADA-SARLABOUS and J.-M. FEUSTEL for their cooperation during several years. I would also like to thank the late KARL-WEIERSTRASS Institute in Berlin and the MAX-PLANCK-Gesellschaft for their support, furthermore U. BURRI and A. MEIER from the University of Basel for writing a manuscript of the first two chapters and M. PFISTER from the ETH Zürich for her helpful technical work on PC.

Berlin, December 1994 Rolf-Peter Holzapfel

1 Elliptic Curves, the Finiteness Theorem of Shafarevič

1.1 Elliptic Curves over $\mathbb{C}$

Instead of giving an introduction we refer to an arithmetic-geometric part of the theory of elliptic curves. Let $\wedge$ be a *lattice in* $\mathbb{C}$, that means a discrete additive subgroup of ($\mathbb{Z}$)-rank 2. Two lattices $\wedge$ and $\wedge'$ in $\mathbb{C}$ are said to be *equivalent*, if there is a complex number $\alpha \neq 0$ such that $\wedge' = \alpha\wedge$. Each of our lattices is equivalent to a lattice $\wedge_\tau = \mathbb{Z} + \mathbb{Z}\tau$ with

$$\tau \in \mathbb{H} = \{z \in \mathbb{C}; \operatorname{Im} z > 0\} .$$

$\mathbb{H}$ is called the POINCARÉ upper half plane. The quotient spaces

$$E_\wedge = \mathbb{C}/\wedge , \quad E_\tau = \mathbb{C}/\wedge_\tau$$

are one-dimensional complex tori, that means complete RIEMANN surfaces with abelian group structures. For equivalent lattices $\wedge$, $\wedge'$ we have a commutative diagram

$$\begin{array}{ccccccccc} 0 & \longrightarrow & \wedge & \longrightarrow & \mathbb{C} & \longrightarrow & E & \longrightarrow & 0 \\ & & \downarrow \wr & & \downarrow \| & & \downarrow \wr & & \\ 0 & \longrightarrow & \wedge' & \longrightarrow & \mathbb{C} & \longrightarrow & E' & \longrightarrow & 0 \end{array}$$

with obvious notations. The tori E, E' are isomorphic. So each $E = E_\wedge$ is isomorphic to a complex torus E_τ for a suitable $\tau \in \mathbb{H}$.

Each torus E has a smooth complex projective algebraic structure. More precisely, it can be analytically embedded into the complex projective plane $\mathbb{P}^2(\mathbb{C})$. A torus together with such an embedding is called an *elliptic curve over* $\mathbb{C}$. For the embeddings we need elliptic functions on $\mathbb{C}$. A meromorphic function on $\mathbb{C}$ is called *elliptic*, if it is $\wedge$-periodic for a suitable $\mathbb{C}$-lattice $\wedge$.

A central role among the elliptic functions play the WEIERSTRASS *$\wp$-functions.* For a fixed lattice $\wedge$ it is defined as

$$\wp_\wedge : \mathbb{C} \longrightarrow \mathbb{P}^1(\mathbb{C}) ,$$

$$\wp_\wedge(z) = 1/z^2 + \sum_{\omega \in \wedge^*} \left(1/(z-\omega)^2 - 1/\omega^2\right) ,$$

where $\wedge^* = \wedge \backslash 0$. The field of meromorphic functions of $E_\wedge$ is generated by $\wp_\wedge$ and $\wp'_\wedge$. Both functions are related by a simple algebraic equation producing a differential equation for $\wp_\wedge$:

$$\wp'_\wedge(z)^2 = 4\wp_\wedge(z)^3 - g_2(\wedge)\wp_\wedge(z) - g_3(\wedge) ,$$

where

$$g_2(\Lambda) = 60 \sum_{\omega\in\Lambda^*} 1/\omega^4 , \quad g_3(\Lambda) = 140 \sum_{\omega\in\Lambda^*} 1/\omega^b .$$

This way we get a projective embedding

$$h : \mathbb{C}/\Lambda \hookrightarrow \mathbb{P}^2(\mathbb{C})$$
$$z \bmod \Lambda \longmapsto (1 : \wp(z) : \wp'(z)) \quad (z \notin \Lambda) .$$

Using projective coordinates $(w : x : y)$ the image curve $= E(\Lambda)$ is defined by the following equation:

$$E : WY^2 = 4X^3 - g_2(\Lambda)W^2X - g_3(\Lambda)W^3 \tag{1.1}$$

Conversely, if E is a smooth projective curve of degree 3, then there is a projectively equivalent curve E' of equation type

$$E' : WY^2 = 4X^3 - g_2W^2X - g_3W^3 . \tag{1.2}$$

Equation (1.2) or the corresponding cubic form is called a *Weierstrass normal form* of E. Moreover, there is a $\mathbb{C}$-lattice Λ such that $g_2 = g_2(\Lambda)$, $g_3 = g_3(\Lambda)$. So we get in any case a *uniformization* $\mathbb{C} \to \mathbb{C}/\Lambda \xrightarrow{\sim} E$.

We want to introduce and to explain now the *moduli space of elliptic curves.*

POINCARÉ'S upper half plane $\mathbb{H}$ is the simplest non-euclidean model of a homogeneous (symmetric) space. On $\mathbb{H}$ the real special linear group $\mathbb{S}l(2,\mathbb{R})$ acts transitively via fractional linear transformations

$$\tau \mapsto (a\tau + b)/(c\tau + d) , \quad \tau \in \mathbb{H} , \quad \begin{pmatrix} a & b \\ c & d \end{pmatrix} \in \mathbb{S}l(2,\mathbb{R}) .$$

The quotient space $\mathbb{S}l(2,\mathbb{Z})\backslash\mathbb{H}$ has a natural complex structure. It is isomorphic to the affine complex line $\mathbb{A}^1(\mathbb{C}) = \mathbb{C}$. Its natural (smooth) compactification is the projective complex line $\mathbb{P}^1(\mathbb{C})$.

This can be made visible by decomposing $\mathbb{H}$ into infinitely many $\mathbb{S}l(2,\mathbb{Z})$-fundamental domains as it has been first done by GAUSS. The elements $S = \left(\begin{smallmatrix} 0 & -1 \\ 1 & 0 \end{smallmatrix}\right)$ and $T = \left(\begin{smallmatrix} 1 & 1 \\ 0 & 1 \end{smallmatrix}\right)$ generate the unimodular group $\mathbb{S}l(2,\mathbb{Z})$. There is a nice central fundamental domain $\mathcal{F}$ as drawn in figure (1.3). By identification of equivalent boundary points one gets $\mathbb{A}^1(\mathbb{C})$ and the compactification by addition of the external boundary point not lying in $\mathbb{H}$. Shifting $\mathcal{F}$ by means of products of S, T, S^{-1}, T^{-1}

one obtains a covering of $\mathbb{H}$ consisting of $\mathbb{S}l_2(2,\mathbb{Z})$-fundamental domains.

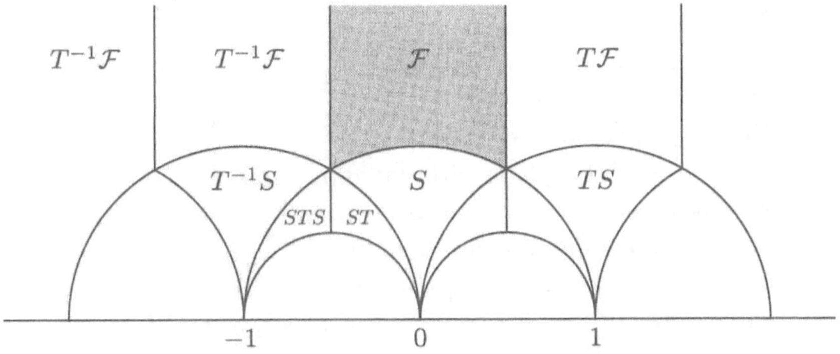

(1.3)

The geometric imagination can be made precise by means of *modular functions.* These are $\mathbb{S}l(2,\mathbb{Z})$-invariant meromorphic functions on $\mathbb{H}$ allowing a meromorphic extension on $\mathbb{S}l(2,\mathbb{Z})\backslash\mathbb{H}$ to the compactification $\mathbb{P}^1(\mathbb{C})$. For $i = 2, 3$ we set $g_i(\tau) = g_i(\wedge_\tau)$. Looking at the discriminant of the polynomial $p_3(X)$ in the WEIERSTRASS equation $Y^2 = p_3(X) = 4X^3 - g_2X - g_3$ of E_τ, we define

$$\Delta(\tau) = 27g_3^2(\tau) - g_2^3(\tau) \ .$$

Then $g_2^3(\tau)/\Delta(\tau)$ is a modular function. The *elliptic modular function* is defined as $j(\tau) = 12^3 g_2^3(\tau)/\Delta(\tau)$. In particular, it is invariant under $S : \tau \mapsto \tau + 1$. It can be written as Fourier series:

$$j(\tau) = q^{-1} + 744q^0 + \sum_{n=1}^{\infty} a_n q^n \,, \quad q = e^{2\pi i \tau} \,, \quad a_n \in \mathbb{Z} \,.$$

The elliptic modular function $j : \mathbb{H} \to \mathbb{C}$ goes down to an analytic isomorphism $\mathbb{S}l(2,\mathbb{Z})\backslash\mathbb{H} \to \mathbb{C}$.

Consider now the elliptic curve family $\mathcal{E}$ over $\mathbb{H}$ defined by

$$\mathcal{E} = \{(w : x : y), \tau) \in \mathbb{P}^2(\mathbb{C}) \times \mathbb{H}; \quad wy^2 = 4x^3 - g_2(\tau)w^2x - g_3(\tau)w^3\}.$$

It has a natural projection onto $\mathbb{H}$. The fibres are the elliptic curves E_τ. The upper half plane $\mathbb{H}$ appears as parameter space for (up to isomorphy) all elliptic curves. This analytic family of curves is denoted by $\mathcal{E}/\mathbb{H}$. The fibres $E_\tau, E_{\tau'}$, are isomorphic iff $\tau' \in \mathbb{S}l(2,\mathbb{Z})\tau$. Therefore we get a bijection

$$\mathbb{C} = \mathbb{S}l(2,\mathbb{Z})\backslash\mathbb{H} \Longleftrightarrow \quad \{\text{isomorphy classes of elliptic curves}\} \,.$$

In this (rough) sense we say that $\mathbb{P}^1$ is the (compactified) *moduli space of elliptic curves.* Altogether we have a commutative diagram for each $\tau \in \mathbb{H}$:

$$\begin{array}{ccccc} E_\tau & \hookrightarrow & \mathcal{E} & \hookrightarrow & \mathbb{P}^2(\mathbb{C}) \times \mathbb{H} \\ \downarrow & & \downarrow & \swarrow & \text{projection} \\ \{\tau\} & \hookrightarrow & \mathbb{H} & & \\ & & \downarrow \mathbb{S}l(2,\mathbb{Z}) & & \\ & & \mathbb{S}l(2,\mathbb{Z}) \backslash \mathbb{H} \cong \mathbb{C} \subset \mathbb{P}^1(\mathbb{C}) & & \end{array}$$

1.2 Elliptic Curves over Arbitrary Fields

We use the following notations:

K	a field, L a field extension of K,
$\bar{K}$	the algebraic closure of K,
$\mathbb{P}^2_K$	the projective plane over K,
$\mathbb{P}^2(L)$	the points of this plane with coordinates in L,
f	a homogeneous polynomial in $K[W, X, Y]$,
$\mathbb{PG}l(3, K)$	the projective linear group $\mathbb{G}l(3, K)/K^*$,
$C : f = 0$	the plane projective curve defined by f,
$C(L)$	the points of C with coordinates in L (L-points).

The group $\mathbb{PG}l(3, L)$ acts on $\mathbb{P}^2(L)$ and $\mathbb{G}l(3, L)$ on $L[W, X, Y]$ in obvious manner. For $G \in \mathbb{G}l(3, L)$ we define the inverse image curve of C by $G^*C : G^*f = 0$, where G^*f denotes the inverse image of f. We have

$$G^*C(L) = \{P \in \mathbb{P}^2(L); G^*f(P) = f(G(P)) = 0\} .$$

Two curves C, C' are called *L-linearly equivalent*, if there is a linear transformation $G \in \mathbb{G}l(3, L)$ such that $C' = G^*C$.

A point $P \in C(L)$ is called *singular* iff the derived polynomials $\partial f/\partial W$, $\partial f/\partial X$, $\partial f/\partial Y$ vanish at P. The curve C is *non-singular* iff each point $P \in C(\bar{K})$ is non-singular.

Definition 1.1. An *elliptic curve* E/K is a non-singular curve of degree 3 in $\mathbb{P}^2_K$ together with a point $0 \in E(K)$.

We are able to define a commutative group structure on E/K. For this purpose consider the L-points of E. Denote by PQ the line through two points $P, Q \in E(L)$. If $P = Q$, then it is defined as tangent line of E through P. By BEZOUT's theorem there is a unique third intersection point $R' \in \mathbb{P}^2(L)$ of $E(\bar{L})$ and $PQ(\bar{L})$ beside of P, Q. It is easy to see that it belongs to $E(L)$. We apply the same procedure to OR' instead of PQ in order to obtain a third intersection point R. Now define $P + Q = R$. Then one gets a commutative group law on $E(L)$, L an arbitrary field extension of K (see [41]). The auxiliary point R' is nothing else than $-(P + Q)$ and O is the neutral element of our addition in figure (1.4).

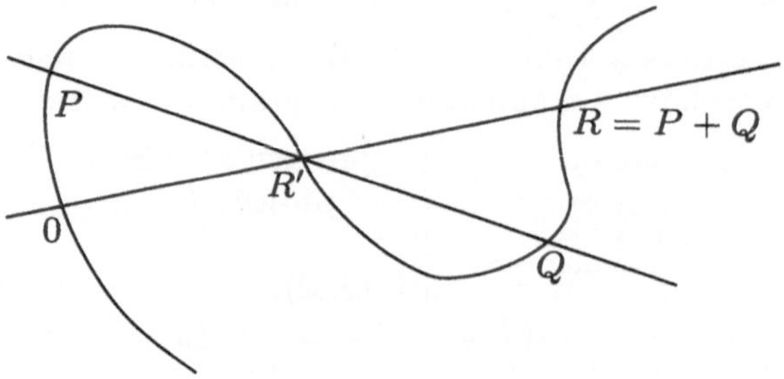

(1.4)

From projective (homogeneous) equations $f = 0$ we change over to affine (inhomogeneous) equations $F = 0$, $F(X,Y) = f(1,X,Y)$. It defines an affine curve in $\mathbb{A}_K^2$ and an affine geometric curve in $\mathbb{A}^2(L)$ as algebraic set of points. Adding some points at infinity ($W = 0$) we get back $C(L)$, especially $C(\bar{L})$, hence $C : f = 0$, $f(W,X,Y) = F(X/W, Y/W)W^{\deg F}$. In our elliptic cases we keep the distinction between affine and projective equations/curves only in mind.

Two elliptic curves E/K, E'/K are *K-(linearly) isomorphic*, iff there exists an element $\alpha \in \mathbb{G}l(3,K)$ such that $E = \alpha^* E'$ and $\alpha(O) = O'$, O' the zero point of E'.

Each elliptic curve E/K is K-isomorphic to an elliptic curve of type

$$E'/K : Y^2 + a_1XY + a_3Y = X^3 + a_2X^2 + a_4X + a_6 \tag{1.5}$$

with $0' = (0:0:1)$, the point at infinity of E'.

If char $K \neq 2,3$, then the above statement remains to be true, if we set $a_i = 0$ for $i = 1,2,3$, that means we substitute (1.5) by

$$E'/K : Y^2 = 4X^3 - g_2X - g_3 \,. \tag{1.6}$$

The equations or curves in (1.5) or (1.6) are called *Weierstrass normal forms* (of E). Up to isomorphy it suffices to investigate elliptic curves given in WEIERSTRASS normal form. So we assume now that

(i) char $K) \neq 2,3$;
(ii) $E/K : Y^2 = 4X^3 - g_2X - g_3$;
(iii) $O = (0:0:1)$; the same for E'/K.

As in the classical (complex) case we look for invariants and their meaning. We set

$$\Delta(E/K) = 27g_3^2 - g_2^3 \,, \quad j(E/K) = 12^3 g_2^3/\Delta(E/K) \,. \tag{1.7}$$

Given a plane projective curve $C/K : f = 0$. We also write C_L, C_L/L or simply C/L for the curve in $\mathbb{P}_L^2$ defined by $f = 0$. With obvious notations and the assumptions (i), (ii), (iii) above, we have the following well-known basic facts:

Proposition 1.2.
(i) *E/K is non-singular, hence an elliptic curve, iff $\Delta(E/K) \neq 0$.*
(ii) *Let E'/L be another elliptic curve, $\bar{L} = \bar{K}$. Then $E/\bar{K}$ and $E'/\bar{K}$ are $\bar{K}$-isomorphic if an only if $j(E/K) = j(E'/L)$ in $\overline{K}$.*
(iii) *The elliptic curves E/K and E'/K are $\bar{K}$-*isomorphic *iff there exists an element $u \in \sqrt{K^\times} = \{v \in \bar{K}; v^2 \in K^\times\}$ such that $g_2' = u^4g_2$, $g_3' = u^6g_3$.*
(iv) *The elliptic curves E/K and E'/K are K-isomorphic iff there exists $u \in K^\times$ such that $g_2' = u^4g$, $g_3' = u^6g_3$.*

1.2.1 Reduction of Elliptic Curves

Let $R \subseteq K$ be an integral domain (with 1), such that $K = \text{Quot } R$, the quotient field of R. We write E/R instead of E/K, if the coefficients of the defining equation belong to R, and we say that E *is defined over* R. An *R-model* of the elliptic curve E'/K is an elliptic curve E/R such that E/K is K-isomorphic to E'/K. It is easy to see that each elliptic curve E'/K has at least one R-model. In fact, there are a lot of them.

Now, let $(R, \mathcal{M})$ be a local ring, $\mathcal{M}$ the maximal ideal of R and $k = R/\mathcal{M}$ the residue field. We write $\bar{g}$ for the residue class of $g \in R$ modulo $\mathcal{M}$. For an elliptic curve $E/R : Y^2 = X^3 - g_2X - g_3$ we define the *reduction* E_k of E/R by

$$E_k/k : Y^2 = X^3 - \bar{g}_2X - \bar{g}_3 \ .$$

We say that E/R has *good reduction*, if E_k is smooth, that is E_k is an elliptic curve over k. There is a nice simple criterion:

Lemma 1.3 (local criterion for good reduction). *The elliptic curve E/R has good reduction if and only if its discriminant $\Delta(E/R)$ is a unit in the local ring R.*

Now let R be a DEDEKIND domain with quotient field $K = \text{Quot } R$, $\mathcal{P} \in \operatorname{Spec} R$ a prime ideal and $R_{\mathcal{P}}$ the corresponding (local) quotient ring. We say that the elliptic curve E'/K has *good reduction at* $\mathcal{P}$, if there is an $R_{\mathcal{P}}$-model $E/R_{\mathcal{P}}$ of E' with good reduction. Otherwise we say that E'/K has *bad reduction at P*. In any case E'/K has good reduction at almost all points of $\operatorname{Spec} R$. If T is a subset of $\operatorname{Spec} R$, then we say that E'/K has *good reduction on* T, if E'/K has good reduction at all points of T. In obvious manner one explains the meaning of *bad reduction outside T*, *bad reduction on $S \subset \operatorname{Spec} R$*, *good reduction outside S*.

In our applications we will work with the ring $R = \mathcal{O}$ of integers of a number field K. Fixing these notations we notice

1.2.2 Two Finiteness Theorems of Number Theory

Denote by $I = I(\mathcal{O})$ the semigroup of integral ideals of $\mathcal{O}$, the group of fractional ideals of K by $I^* = I^*(\mathcal{O}) = I^*(K)$ and by $H^* = H^*(K)$ its subgroup of principal ideals. The group $Cl(K) = I^*/H^*$ is called the *class group of K*.

Theorem 1.4 (Finiteness of class group). *The class group $Cl(K)$ has finite order.*

The order $h(K) = \sharp Cl(K)$ is called the *class number of K*.

For a subset $S \subseteq \operatorname{Spec} \mathcal{O}$ the *ring of S-integers* of K is defined by

$$\mathcal{O}_S = \{a/b;\ a, b \in \mathcal{O},\ b \notin \mathcal{P} \text{ for all } \mathcal{P} \in T = \operatorname{Spec} \mathcal{O} \backslash S\} \ .$$

Note the difference between the local ring

$$\mathcal{O}_{\mathcal{P}} = \{a/b;\ a, b \in \mathcal{O},\ b \notin \mathcal{P}\}$$

and the global ring $\mathcal{O}_{\{\mathcal{P}\}}$.

Corollary 1.5. *For each finite $S' \subset \operatorname{Spec}\mathcal{O}$ there exists a finite $S \subset \operatorname{Spec}\mathcal{O}$ containing S' such that $\mathcal{O}_S$ is a principal domain.*

Proof: The semigroup homomorphism

$$I(\mathcal{O}) \longrightarrow I(\mathcal{O}_S)\,,\ \mathcal{A} \longmapsto \mathcal{A}_S = \mathcal{O}_S\mathcal{A}$$

extends to the exact sequence of group homomorphisms

$$1 \longrightarrow \langle S\rangle \longrightarrow I^*(\mathcal{O}) \longrightarrow I^*(\mathcal{O}_S)\,, \tag{1.8}$$

where $\langle S\rangle$ denotes the group generated by S.

Now let $\{\mathcal{A}_1,\dots,\mathcal{A}_h\}$ be a system of representatives of the class group $Cl(\mathcal{O})$ and

$$S = S' \cup \{\text{prime divisors of } \mathcal{A}_1 \cdot \ldots \cdot \mathcal{A}_h\}\,.$$

For each ideal $\mathcal{A}$ of K we find $a \in K$ and $i \in \{1,\dots,h\}$ such that $\mathcal{A}_S = (a\mathcal{A}_i)_S = a\mathcal{O}_S$ because of $\mathcal{A}_i \in \langle S\rangle$ and (1.8). □

Theorem 1.6 (DIRICHLET's Unit Theorem). *For finite $S \subset \operatorname{Spec}\mathcal{O}$ the group of units $\mathcal{O}_S^*$ of $\mathcal{O}_S$ is finitely generated.*

Corollary 1.7. *For each natural number n the factor group $\mathcal{O}_S^*/\mathcal{O}_S^{*n}$ is finite.*

1.2.3 SHAFAREVIČ's Finiteness Theorem

Lemma 1.8 (global criterion for good reduction). *Let S be a finite subset of $\operatorname{Spec}\mathcal{O}_S$ such that $\mathcal{O}_S$ is a principal domain. The elliptic curve E'/K has good reduction outside of S iff it has an $\mathcal{O}_S$-model $E/\mathcal{O}_S$ such that $\Delta(E/\mathcal{O}_S) \in \mathcal{O}_S^*$.*

Proof: The discriminant condition is sufficient by the local criterion 1.3.

Assume conversely that for each $\mathcal{P} \in T = \operatorname{Spec}\mathcal{O}\backslash S$ there is a model

$$E_{\mathcal{P}}/\mathcal{O}_{\mathcal{P}} : Y^2 = 4X^3 - g_{2\mathcal{P}}X - g_{3\mathcal{P}}$$

of E'/K with $\Delta_{\mathcal{P}} = \Delta(E_{\mathcal{P}}/\mathcal{O}_{\mathcal{P}}) \in \mathcal{O}_{\mathcal{P}}^*$. With obvious notations we have

$$g_2' = u_{\mathcal{P}}^4 \cdot g_{2\mathcal{P}}\,,\ g_3' = u_{\mathcal{P}}^6 \cdot g_{3\mathcal{P}}, \quad \Delta' = u_{\mathcal{P}}^{12}\Delta_{\mathcal{P}} \tag{1.9}$$

for suitable $u_{\mathcal{P}} \in K$, $\mathcal{P} \in T$. Without loss of generality we can assume that we start with a model $E'/\mathcal{O}_K$, hence $g_i' \in \mathcal{O}_K$. Let $\{\mathcal{P}_1,\dots,\mathcal{P}_r\}$ be the set of prime divisors of $\Delta' \in \mathcal{O}_K$. Then

$$u_{\mathcal{P}} \in \mathcal{O}_{\mathcal{P}}^* \quad \text{for} \quad \mathcal{P} \in T\backslash\{\mathcal{P}_1,\dots m\mathcal{P}_r\}$$

by the last identities of (1.9) and our assumptions. So $(\mathcal{O}_\mathcal{P} u_\mathcal{P})_{\mathcal{P}\in T}$ belongs to the restricted product group (with components 1 almost everywhere)

$$\prod_{\mathcal{P}\in T}{}' I^*(\mathcal{O}_\mathcal{P}) \xrightarrow{\sim} I^*(\mathcal{O}_S)\,.$$

Since $\mathcal{O}_S$ is principal we can represent our tuple by $\mathcal{O}_S u$, $u \in K$; so

$$u_\mathcal{P} = \varepsilon_\mathcal{P} u\,,\ \varepsilon_\mathcal{P} \in \mathcal{O}_\mathcal{P}^* \quad \text{for all} \quad \mathcal{P} \in T\,. \tag{1.10}$$

Now we define the elliptic curve

$$E/\mathcal{O}_S : Y^2 = X^3 - g_2 X - g_3$$

setting

$$g_2 = g_2'/u^4\,,\ g_3 = g_3'/u^6\,. \tag{1.11}$$

The coefficients of the equation of E differ from those of $E_\mathcal{P}$ only by local units because of (1.11), (1.9) and (1.10). This is also true for $\Delta = \Delta(E/\mathcal{O}_S)$ and Δ' for the same reasons. Therefore $\Delta \in \mathcal{O}_\mathcal{P}^*$ for all $\mathcal{P} \in T$, hence $\Delta \in \mathcal{O}_S^*$. □

Theorem 1.9 (SHAFAREVIČ). *Let K be a number field, $\mathcal{O} = \mathcal{O}_K$ its ring of integers and S a finite set of prime ideals of $\mathcal{O}$. Then, up to K-*isomorphy, *there are only finitely many elliptic curves E/K with good reduction outside of S.*

Proof: Without loss of generality we can assume that all prime divisors of 2 and 3 belong to S. So we can work locally along $T = \operatorname{Spec}\mathcal{O} \setminus S$ and also globally with WEIERSTRASS normal forms in the narrow sense of (1.6). The class of all elliptic curves E/K with good reduction outside of S is denoted by $\mathcal{E}(K,S)$. The domain can be assumed to be principal by Corollary 1.5. Each member of $\mathcal{E}(K,S)$ has models $E/\mathcal{O}_S$ with $\Delta(E/\mathcal{O}_S) \in \mathcal{O}_S^*$ by Lemma 1.8. Together with Proposition 1.2 (iv) we see that the map

$$\delta : \mathcal{E}(K,S) \longrightarrow \mathcal{O}_S^*/\mathcal{O}_S^{*12}\,,\ E/\mathcal{O}_S \longmapsto \Delta(E/\mathcal{O}_S)\mathrm{mod}^\times \mathcal{O}_S^{*12}$$

is well-defined. The image is finite by Corollary 1.7. So it suffices to prove that for a given S-unit D there exist only finitely many elliptic curves

$$E/\mathcal{O}_S : Y^2 = X^3 - g_2 X - g_3$$

with $\Delta(E/\mathcal{O}_S) = D$. This follows immediately from the definition of the discriminant and the next lemma. □

Lemma 1.10. *With the above notations the diophantine equation*

$$U^3 - 27V^2 = D$$

has only finitely many solutions u, v *in* $\mathcal{O}_S$.

1.2.4 Basic References

For an introduction to the classical theory of elliptic and modular functions we refer to [46]. All we need in 1.1 can be found in the first chapters there. The omitted proofs of some basic results on elliptic curves over finite fields are contained in [41]. K-isomorphy of curves needs in general the finer scheme language. It will be necessarily used later. Our style of writing is a good preparation. The basic introduction is HARTSHORNE's book [27]. Proofs of the two basic finiteness theorems 1.4 and 1.6 can be found in [16].

Our proof of SHAFAREVIČ's Finiteness Theorem for elliptic curves is a detailed version of SERRE's proof in [69]. The theorem was announced by SHAFAREVIČ of the International Congress in Stockholm 1962, together with a far-reaching conjecture on algebraic curves over number fields (SHAFAREVIČ conjecture) proved by FALTINGS in 1983 together with the MORDELL conjecture as consequence. The diophantine equation in Lemma 1.10 can be solved effectively by methods of BAKER [4], see also SERRE's lectures [71]. Altogether one has an effective way of finding up to isomorphy all elliptic curves over a fixed number field with prescribed places of bad reduction. An algorithm has been established by TATE [88].

Recently ESTRADA-SARLABOUS found a way to transfer the methods and the effective result to PICARD curves

$$C : Y^3 = X^4 + G_2X^2 + G_3X + G_4$$

of genus 3 (see Appendix I). These curves play a central role in all the following chapters.

2 Picard Curves

2.1 The Moduli Space of Picard Curves

Definition 2.1. Let C' be a compact algebraic curve over $\mathbb{C}$. It is called a *Picard curve*, if it is isomorphic to a plane projective curve $C/\mathbb{C}$ of the following equation type:

$$C' \xrightarrow{\sim} C : WY^3 = \sum_{i=0}^{4} G_i W^i X^{4-i}, \quad G_0 \neq 0$$

In affine coordinates the plane Picard curve C is described by

$$C : Y^3 = G_0X^4 + G_1X^3 + G_2X^2 + G_3X + G_4 \ .$$

One has to add the point $\infty = (0 : 0 : 1)$ in order to obtain the projective model from the affine one. By means of projective Tschirnhaus transformation one can reduce the equations to the following *normal forms*

$$\begin{aligned} WY^3 &= X^4 + G_2W^2X^2 + G_3W^3X + G_4W^4 \text{ (projective)}, \\ Y^3 &= X^4 + G_2X^2 + G_3X + G_4 = p_4(X) \text{ (affine)}. \end{aligned} \tag{2.1}$$

The singular locus of

$$C : F(W, X, Y) = WY^3 - X^4 - G_2W^2X^2 - G_3W^3X - G_4W = 0$$

can be determined by solving the system of homogeneous equations

$$F = \partial F/\partial W = \partial F/\partial X = \partial F/\partial Y = 0 \ . \tag{2.2}$$

The point ∞ is a smooth one because $\partial F/\partial W(0, 0, 1) = 1$. So all singular points of C lie in the affine part. It is easy to see that only the intersection points with the line $L_0 : Y = 0$ are possible singularities. These are the points

$$R_i = (1 : a_i : 0) \ , \quad i = 1, \dots, 4 \ , \tag{2.3}$$

where $a_1, \dots, a_4$ are the zeros of $p_4(X)$. As in the case of elliptic curves we have a *discriminant* criterion: $\Delta(C) \neq 0$. The discriminant of C is defined as $\Delta(C) = \prod_{i \neq j} (a_j - a_i)$. In terms of the coefficients of F it is described by

$$\Delta(C) = 16G_2^4 \cdot G_4 - 128G_2^2 \cdot G_4^2 - 4G_2^3 \cdot G_3^2 + 144G_2G_3^2G_4 - 27G_3^4 + 256G_4^3.$$

Picture (2.4) gives an imagination of (the real part of) a PICARD curve in normal form with exactly one (real) singularity.

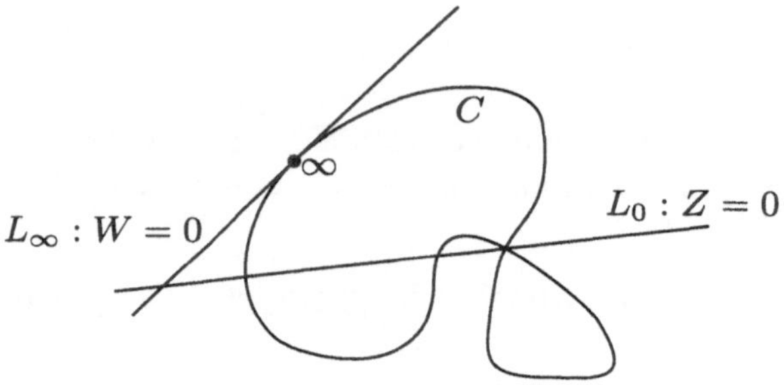

(2.4)

The line L_∞ touches C at ∞ of order (intersection number) 4.

We look now for the moduli space $\mathbf{M}$ of PICARD curves in the rough sense: to find a complex-algebraic structure on the set of isomorphy classes of these curves. More precisely, this will be done for smooth curves, and then we look for a natural compactification and interpretation:

$$\{\text{smooth PICARD curves}\}/\text{Isom.} \Longleftrightarrow \mathbf{M}^0 \subset \mathbf{M}$$

Set

$$\mathbb{C}_0^4 = \{(z_1, \ldots, z_4) \in \mathbb{C}^4;\ z_1 + \ldots + z_4 = 0\} \subset \mathbb{C}^4$$

and let $\mathcal{C}$ be the following analytic family of PICARD curves:

$$\mathcal{C} = \left\{((w : x : y), (a_1, \ldots, a_4)) \in \mathbb{P}^2(\mathbb{C}) \times \mathbb{C}_0^4;\ wy^3 = \prod_{i=1}^{4}(x - a_i w)\right\}$$

Without change of the notation $\mathcal{C}$ we omit the special singular fibre with $WY^3 = X^4$ over 0. All other PICARD curves are represented in $\mathcal{C}$ up to isomorphy. We have the following commutative diagrams

$$\begin{array}{ccccccc} C_a & \hookrightarrow & \mathcal{C} & \hookrightarrow & \mathbb{P}^2 & \times & \mathbb{C}_0^4 \\ \downarrow & & \downarrow & \swarrow & & & \downarrow \\ \{a\} & \hookrightarrow & \mathbb{C}_0^4 \setminus 0 & \longrightarrow & & & \mathbb{P}\mathbb{C}_0^4 = \mathbb{P}\mathbb{C}^3 = \mathbb{P}^2 \end{array} \qquad (2.5)$$

with obvious projections and identifications.

The symmetric group S_4 acts on $\mathbb{C}_0^4$ by permutation of coordinates. This action goes down to $\mathbb{P}^2$. The compact quotient surface $\hat{\mathbf{M}} = \mathbb{P}^2/S_4$ is normal, algebraic and, by LÜROTH's theorem, rational.

We go back to $\mathbb{P}^2 = \mathbb{P}_0^3 := \mathbb{P}\mathbb{C}_0^4$ writing the elements as homogeneous quadruples $(a_1 : \ldots : a_4)$, $a_1 + \ldots + a_4 = 0$. Now we choose four points in general position. To be explicit we choose

$$\begin{aligned} P_1 &= (-3 : 1 : 1 : 1), \quad P_2 = (1 : -3 : 1 : 1)\,, \\ P_3 &= (1 : 1 : -3 : 1), \quad P_4 = (1 : 1 : 1 : -3)\,. \end{aligned} \qquad (2.6)$$

The line through P_i, P_j is denoted by $L_{ij} = L_{ji}$. These six lines form a reduced divisor

$$\triangle\!\!\!\!\star = L_{12} + L_{13} + L_{14} + L_{23} + L_{24} + L_{34} \tag{2.7}$$

on $\mathbb{P}^2$ as described in picture (2.8)

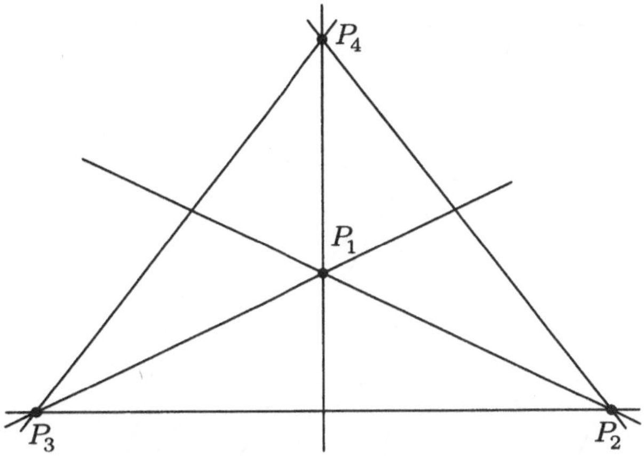

(2.8)

Obviously the action of the symmetric group S_4 restricts to an action on $\mathbb{P}^2 \setminus \Delta$. We set

$$\mathbf{M}^0 := \left(\mathbb{P}^2 \setminus \triangle\!\!\!\!\star\right)/S_4 \subset \mathbf{M} := \mathbb{P}^2 \setminus \{P_1, \ldots, P_4\} \subset \hat{\mathbf{M}} := \mathbb{P}^2/S_4 \; .$$

Two plane PICARD curves C, C' are called *linearly isomorphic*, if there is a $G \in Gl_3(\mathbb{C})$ such that $G^*C = C'$.

Proposition 2.2. *Two* PICARD *curves* $C_a, C_{a'} (a, a' \in \mathbb{C}_0^4 \setminus 0)$ *are linearly isomorphic if and only if* $\mathbb{P}a$ *and* $\mathbb{P}a'$ *are* S_4-equivalent *points of* $\mathbb{P}^2$.

Idea of proof. An old classical invariant is useful. The *Hesse form* of a homogeneous polynomial $F = F(X_0, \ldots, X_n)$ is defined as

$$\mathrm{Hess}(F) = \det(\partial^2 F/\partial X_i \partial X_j) \; .$$

Geometrically it corresponds to each effective divisor $D : F = 0$ on $\mathbb{P}^n$ an effective divisor called the *Hessian* of D (or F):

$$\begin{aligned} \mathrm{Hess} : \mathrm{Div}^+\mathbb{P}^n &\longrightarrow \mathrm{Div}^+\mathbb{P}^n \\ D &\longmapsto \mathrm{hess}\,(D) : \mathrm{Hess}\,(F) = 0 \end{aligned}$$

The Hessian is a projective invariant of divisors. This means that it does not change under projective transformations:

$$\mathrm{hess}\,(G^*D) = \mathrm{hess}\,(D) \; , \quad D \in \mathrm{Div}^+\mathbb{P}^n \; , \quad G \in \mathbb{G}l(n+1, \mathbb{C})$$

Now we define a special invariant of PICARD curves $C : F = 0$, with normal form F as used in (2.1). Let L_i be the line through R_i (see (2.3)) and ∞, $i = 1, \dots, 4$. The divisor

$$R(C) = L_0 + L_\infty + L_1 + L_2 + L_3 + L_4 \in \mathrm{Div}^1\mathbb{P}^2$$

is called the *rack* of C. It is immediately clear that the rack $R(C)$ determines uniquely the PICARD normal form of C because the knowledge of $R_1, \dots, R_4$ is sufficient, see picture (2.9).

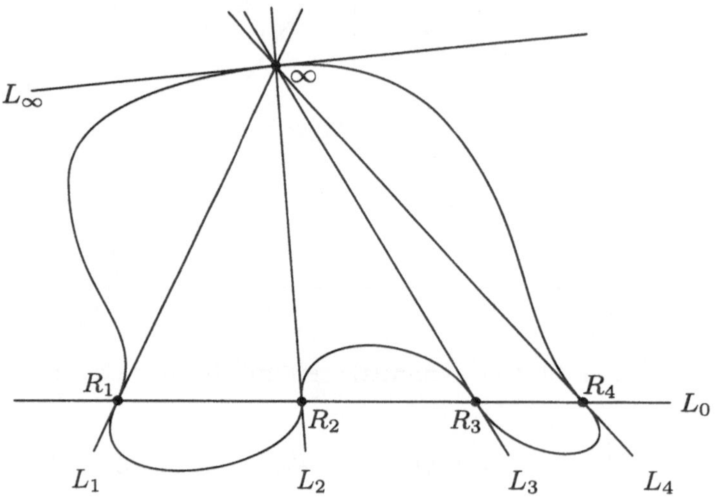

(2.9)

$$\begin{aligned} L_\infty &: & W &= 0 \\ C &: & WY^3 &= \prod_{i=1}^{4}(X - a_i X) \\ & & &= X_4 + G_2W^2X^2 + G_3W^3X + G_4W \\ L_0 &: & Y &= 0 \end{aligned}$$

Now we conclude as follows in the general case $G_2, G_3 \neq 0$, C smooth. The other special cases are left to the reader (see [33]).

A simple calculation shows that L_0 is the only irreducible component of degree 1 of hess (C). Therefore it is intrinsically defined by C together with the intersection points $R_1, \dots, R_4$ of L_0 and C. The tangent lines of C at R_i are the lines $L_i, i = 1, \dots, 4$. They meet each other at ∞. Hence also L_∞ and finally $R(C)$ can be recovered from the projective invariant hess (C). In this way the definition of the rack $R(C)$ extends to all plane PICARD curves not necessarily given in normal form.

Moreover, for each linear transformation G and each plane PICARD curve C it holds that

$$G^*R(C) = R(G^*C)\ , \quad G \in \mathbb{G}l(3, \mathbb{C})\ . \tag{2.10}$$

Now we assume that the PICARD curves C, C', both given in normal form, are linearly equivalent, say $C' = G^*C$. By (2.10) and (2.9) we get $G^*L_\infty = L_\infty$, $G^*L_0 = L_0$. With respect to the equations, G is a TSCHIRNHAUS transformation. It induces a permutation of the coordinates of $(a_1 : \ldots : a_4)$ because our equations are assumed to be normal. □

Proposition 2.3. *The moduli space of smooth* PICARD *curves is*

$$\mathbf{M}^0 = \left(\mathbb{P}^2 \setminus \triangle\!\!\!\!\star\,\right)/S_4.$$

Proof: Let C be a smooth PICARD curve. It is isomorphic to a smooth projective plane curve of degree $d = 4$ by definition. By a well-known genus formula we can calculate

$$g(C) = (d-1)(d-2)/2 = 3\ . \tag{2.11}$$

For smooth plane curves of genus 3 there is no difference between linear isomorphy and isomorphy by the next lemma. □

Lemma 2.4. *For two smooth plane projective curves C, C' of genus 3 (of degree 4) it holds that $C \cong C'$ iff C and C' are linearly isomorphic.*

Proof: Let C be an arbitrary smooth complete algebraic curve of genus 3 and $\omega = \omega_C$ the *canonical sheaf* (of regular differential forms) on C. The $\mathbb{C}$-vector space $H^0(C, \omega)$ has dimension 3 and the *canonical map*

$$\phi_\omega : C \longrightarrow \mathbb{P}^2\,, \qquad P \longmapsto (w_0(P) : w_1(P) : w_2(P))\ ,$$

$\{w_0, w_1, w_2\}$ a basis of $H^0(C, \omega)$, is a closed embedding. It is uniquely determined up to projective linear equivalence. So the corresponding images are linearly isomorphic curves of degree 4 by the degree formula (2.11). If C' is another curve isomorphic to C, then, by suitable choices of basis, we can arrange that the canonical images coincide. It follows that $\phi_\omega(C)$ and $\phi_{\omega'}(C)$ are linearly isomorphic. It remains to check that the canonical map ϕ_ω itself is a linear isomorphy, if C is a smooth plane projective curve of genus 3. This follows from the adjunction formula

$$\omega_C \cong \omega_{\mathbb{P}^2} \otimes \mathcal{O}(C) \cong \mathcal{O}(-3) \otimes \mathcal{O}(4) \mid_C = \mathcal{O}(1) \mid_C,$$

where $\mathcal{O} = \mathcal{O}_{\mathbb{P}^2}$ is the structure sheaf of $\mathbb{P}^2$, $\mathcal{O}(1) = \mathcal{O}(H)$, H a projective line, and $\mathcal{O}(n) = \mathcal{O}(1)^n$. The lemma is proved.

In order to finish the proof of Proposition 2.3 we establish the bijection

$$\{\text{smooth PICARD curve}\}/\text{Isom.} \underset{m}{\Longleftrightarrow} \mathbf{M}^0 \subset \mathbf{M} \tag{2.12}$$

we look for as follows:

$$C \mapsto \phi_\omega(C) \mapsto C_a \mapsto (a_1 : a_2 : a_3 : a_4)S_4 = m(C)$$

C is the smooth PICARD curve we start with. Its canonical image is a plane curve of degree 4 and C_a a PICARD normal form. The normal form is uniquely determined up to linear isomorphy by Lemma 2.4.

Finally, the *moduli point* $(a_1 : \ldots : a_4)S_4$ is unique by Proposition 2.2.

By the discriminant criterion C is smooth if and only if $a_1, \ldots, a_4$ are four different numbers. This means that the moduli point $m(C)$ belongs to $\left(\mathbb{P}^2 \setminus \triangle\!\!\!\!\star\right)/S_4$.
□

By definition of PICARD curves we have an analytic 3- parametric family $\mathcal{C}/\mathbb{C}_0^4$ of PICARD curves containing all of them as fibres up to isomorphy (look before (2.5)). Since the moduli space above has dimension 2, we look for an algebraic 2-parameter family with the same property of completeness. Unfortunately we cannot use the moduli space $\mathbf{M}^0$ (or $\mathbf{M}, \hat{\mathbf{M}}$) itself as basic space. But we can do it by introducing a level structure on PICARD curves.

An *ordered* PICARD curve in normal form is a PICARD curve $C_a : Y^3 = \prod_{i=1}^{4}(X - a_i)$ together with a bijective correspondence

$$\alpha : \{1,2,3,4\} \longrightarrow \{R_1, R_2, R_3, R_4\}\ , \tag{2.13}$$

where the points R_i are defined in (2.3). The definition extends correctly to arbitrary smooth PICARD curves C along the moduli map (2.12). Namely, the canonical model $C' \subset \mathbb{P}^2$ of C is unique up to linear isomorphy, and the points $R_1, \ldots, R_4$ can be read off from the rack $R(C')$. From (2.10) follows that the preimage $\{R_1, \ldots, R_4\} \subset C$ of $\{R'_1, \ldots, R'_4\} \subset C'$ along the canonical map $C \to C'$ does not depend on the choice of the canonical model. Therefore the pair $(C, \{R_1, \ldots, R_e\})$ is well-defined and also the (smooth) ordered PICARD curve (C, α) with α as in (2.13).

Two ordered PICARD curves (C, α) and (C', α') are called isomorphic, if there is an isomorphism $f : C \xrightarrow{\sim} C'$ such that $f \circ \alpha = \alpha'$. From (2.12) it is immediately clear that we have a bijective map

$$\{\text{smooth ordered PICARD curves}\}/\text{Isom.} \Longleftrightarrow \mathbb{P}^2 \setminus \triangle\!\!\!\!\star \tag{2.14}$$

Definition 2.5. We call $\mathbb{P}^2 \setminus \triangle\!\!\!\!\star$ the *moduli space of smooth ordered* PICARD *curves.* The open quotient surface $\mathbf{M}^0 \subset \mathbb{P}^2/S_4$ is called the *moduli space of smooth* PICARD *curves.*

With the notations of (2.6) we call sometimes $\mathbb{P}^2\setminus\{P_1,\ldots,P_4\}$ or $\mathbb{P}^2$ the *moduli space of ordered* PICARD *curves* respectively the compact(ified) moduli space of ordered PICARD curves and $(\mathbb{P}^2\setminus\{P_1,\ldots,P_4\})/S_4$ or $\mathbb{P}^2/S_4$ the *moduli space of* PICARD *curves* respectively the compact(ified) moduli space of PICARD curves.

Now we can construct an algebraic family $\mathcal{C}/(\mathbb{P}^2 \setminus \triangle)$ of (smooth ordered) PICARD curves containing all smooth PICARD curves as fibres up to isomorphy. For this purpose we define the affine parameter space $T \subset \mathbb{C}^2$ as the complement in $\mathbb{C}^2$ of the five lines

$$s=0\,,\ s=1\,,\ t=0\,,\ t=1\,,\ s=t\,,$$

using coordinates (s,t) on $\mathbb{C}^2$. It is clear that $T \cong \mathbb{P}^2 \setminus \triangle$. The PICARD *family* $\mathcal{C}/T \subset \mathbb{P}^2 \times T/T$ is defined by

$$\mathcal{C} = \left\{((w:x:y),(t_1,t_2)) \in \mathbb{P}^2 \times T\,;\ wy^3 = x(x-w)(x-t_1w)(x-t_2w)\right\}\ . \tag{2.15}$$

2.2 The Relative SCHOTTKY Problem for PICARD Curves

Let C be a compact RIEMANN surface of genus g, $\mathcal{O} = \mathcal{O}_C$ the sheaf of holomorphic functions and $\omega = \omega_C$ the sheaf of holomorphic differential forms (type $(1,0)$) on C. By SERRE duality the cohomology groups $H^1(C,\mathcal{O})$ and $H^0(C,\omega)$ are isomorphic. For abbreviation we denote the latter $\mathbb{C}$-vector space by H^1 and the homology group $H_1(C,\mathbb{Z})$ by H_1 or $H_1(\mathbb{Z})$. We know that

$$\dim H^1 = g\,,\quad rk\,H_1 = 2g\,.$$

The integral map

$$\int : H_1 \times H^1 \to \mathbb{C}\,,\quad (\alpha,\omega) \mapsto \int_\alpha \omega \tag{2.16}$$

is bilinear in both arguments. The integral $\int_\alpha \omega$ is called the *period integral* of ω along α.

Let $\vec{\omega} = {}^t(\omega_1,\ldots,\omega_g)$ or $B = (\alpha_1,\ldots,\alpha_{2g})$ be a basis of H^1 or H_1, respectively. The column

$$\int_\alpha \vec{\omega} = {}^t\!\left(\int_\alpha \omega_1,\ldots,\int_\alpha \omega_g\right) \in \mathbb{C}^g \tag{2.17}$$

is called a *period vector*. The matrix

$$\Pi = \Pi(\vec{\omega},B) = \int_B \vec{\omega} = \left(\int_{\alpha_1} \vec{\omega},\ldots,\int_{\alpha_{2g}} \vec{\omega}\right) \in \mathrm{Mat}_{g,2g}(\mathbb{C})\ , \tag{2.18}$$

is called a *generalized period matrix* and

$$\Lambda_\Pi = \Lambda(\vec{\omega}, B) = \mathbb{Z}\int_{\alpha_1}\vec{\omega} + \ldots + \mathbb{Z}\int_{\alpha_{2g}}\vec{\omega} \subset \mathbb{C}^g \tag{2.19}$$

a *period lattice.*

The groups $\mathbb{G}l_g(\mathbb{C})$ or $\mathbb{G}l_{2g}(\mathbb{Z})$ act from the left or from the right-hand side on the set of generalized period matrices induced by base changes in H^1 or $H_1(\mathbb{Z})$, respectively:

$$\mathbb{G}l_g(\mathbb{C}) : \circlearrowleft \{\text{ generalized period matrices }\} \circlearrowright : \mathbb{G}l_{2g}(\mathbb{Z}) \tag{2.20}$$

The homology group $H_1(\mathbb{Z})$ is endowed with an intersection product $\circ : H_1 \times H_1 \longrightarrow \mathbb{Z}$. This is an integral skew-symmetric unimodular bilinear form. In other words: $(H_1, \circ)$ is a *symplectic $\mathbb{Z}$-module.* By a theorem of FROBENIUS there exists a *symplectic* (or *normal*) *basis* $(\alpha_1, \ldots . \alpha_{2g})$ of H_1 defined by

$$(\alpha_i \circ \alpha_j) = I = I_{2g} = \left(\begin{array}{c|c} 0 & E_g \\ \hline -E_g & 0 \end{array}\right) . \tag{2.21}$$

A *period matrix* is a generalized period matrix (2.18) with a symplectic basis B. In analogy to (2.20) the *symplectic modular group* defined by

$$\mathbb{S}p(2g, \mathbb{Z}) = \{g \in \mathbb{G}l_{2g}(\mathbb{Z})\,;\ {}^t gIg = I\} \tag{2.22}$$

acts from the right on the set of period matrices of C.

$$\mathbb{G}l_g(\mathbb{C}) : \circlearrowleft \{\text{ period matrices }\} \circlearrowright : \mathbb{S}p(2g,\mathbb{Z}) \tag{2.23}$$

In the following text we will only work with symplectic bases, hence with period lattices and period matrices in this stronger sense.

The *Jacobian (variety) of C* is the compact complex commutative Lie group $J(C) = \mathbb{C}^g/\Lambda$, Λ a period lattice. It is uniquely defined by C up to isomorphy. Namely, for a fixed basis $\vec{\omega}$ of H^1 the period lattice $\Lambda = \Lambda(\vec{\omega})$ does not depend on the choice of the symplectic basis by definition (2.19) and (2.23). If we use another basis $\vec{\omega}' = g\vec{\omega}$ of H^1 and set $\Lambda' = \Lambda(\vec{\omega}')$, $J' = \mathbb{C}^g/\Lambda'$, then we have a commutative diagram

$$\begin{array}{ccccccccc} 0 & \longrightarrow & \Lambda & \longrightarrow & \mathbb{C}^g & \longrightarrow & J & \longrightarrow & 0 \\ & & \downarrow\wr & & \downarrow\wr & & \downarrow\wr & & \\ 0 & \longrightarrow & \Lambda' & \longrightarrow & \mathbb{C}^g & \longrightarrow & J' & \longrightarrow & 0 \end{array}$$

with obvious notations.

2.2.1 The JACOBI Map

Fix a point 0 on C and a H^1-basis $\vec{\omega}$. The JACOBI *map* (with origin 0) is defined by

$$j : C \longrightarrow J(C)\,, \quad P \longmapsto \int_0^P \vec{\omega} \bmod \wedge\,. \tag{2.24}$$

This is an analytic embedding. Depending on the precision we need it is also denoted by j_C, j_0 or $j_{C,0}$.

Let $\mathrm{Div}(C)$, $\mathrm{Div}^0(C)$, $\mathrm{Div}^h(C)$ be the group of divisors on C, its subgroup of divisors of degree 0 or its subgroup of principal divisors, respectively. $\mathrm{Div}^0(C)$ is generated by the special divisors $P - 0$, $P \in C$. The JACOBI map j extends to

$$j' : \mathrm{Div}^0(C) \longrightarrow J(C) \tag{2.25}$$

$$\Sigma m_p \cdot (P - 0) \longmapsto \left(\Sigma m_p \cdot \int_0^P \vec{\omega} \right) \bmod \wedge\,.$$

By the theorems of ABEL and JACOBI the sequence

$$0 \longrightarrow \mathrm{Div}^h(C) \longrightarrow \mathrm{Div}^0(C) \underset{j'}{\longrightarrow} J(C) \longrightarrow 0$$

is exact.

2.2.2 RIEMANN's Period Relations

An *abelian variety* over $\mathbb{C}$ is a (smooth) complex projective variety with a commutative (algebraic) group structure. Forgetting the algebraic structure, each abelian variety is a complex torus. But not each *complex torus* $\mathbb{C}^g/\wedge$, $\wedge$ a lattice in $\mathbb{C}^g$, is an abelian variety. Let us consider a set of free generators of $\wedge$ as columns of a matrix $\Pi \in \mathrm{Mat}_{g,2g}(\mathbb{C})$. With these notations one has the following characterization by RIEMANN's

Theorem 2.6. *The complex torus $\mathbb{C}^g/\wedge$ is an abelian variety if and only if*

$$\Pi I^t \Pi = 0\,, \quad i \Pi I^t \bar{\Pi} > 0\,. \tag{2.26}$$

Here "> 0" means that the matrix is positive definite, $i = \sqrt{-1}$, and I is defined in (2.21).

The relations in (2.26) are called the *first* or *second* RIEMANN *(period) relations*, respectively. The last and the next theorem, due to RIEMANN too, show that the Jacobian varieties $J(C)$ of curves are abelian varieties.

Theorem 2.7. *If $\wedge$ is the period matrix of a curve C, then the* RIEMANN *relations* (2.26) *are satisfied.*

Recall the $\mathbb{G}l_g(\mathbb{C})$-action on the period matrices of a fixed curve of genus g, see (2.23). According to this action we can normalize our period matrices in the following manner. Divide a period matrix into two quadratic parts, $\Pi = (\Pi_1|\Pi_2)$ and set $G = \Pi_1^{-1}$. We call $(E_g|\Omega) = G\Pi = G(\Pi_1|\Pi_2)$ a *normalized period matrix*, Ω a *period point*. By (2.23) Ω is uniquely determined by the curve C we work on, up to $\mathbb{S}p(2g,\mathbb{Z})$-equivalence. RIEMANN's period relations transfer to period points as

$$\Omega = {}^t\Omega\,, \quad Im\,\Omega > 0\,. \tag{2.27}$$

The relations give rise to introduce the following analytic space playing an important role in the theory of abelian varieties.

Definition 2.8. The domain

$$\mathbb{H}_g = \{\Omega \in \mathbb{G}l_g(\mathbb{C})\,;\ \Omega = {}^t\Omega\,,\ Im\,\Omega > 0\}$$

is called *generalized upper half plane* or SIEGEL*'s upper half space.*

In analogy to (2.23) the modular symplectic group $\mathbb{S}p(2g,\mathbb{Z})$ acts on $\mathbb{H}_g$. Moreover, $\mathbb{H}_g$ is a symmetric domain with the real symplectic Lie group $Sp(2g,\mathbb{R})$ acting transitively on it. Notice that for a curve C as above

2.9. *the moduli point Ω mod $\mathbb{S}p(2g,\mathbb{Z})$ of C on the algebraic quotient variety $\mathbb{H}_g/Sp(2g,\mathbb{Z})$ is uniquely determined by C.*

Now we are able to formulate the SCHOTTKY problem in a convenient manner: *Find a criterion, which allows to decide whether a matrix $\Pi \in Mat_{g,2g}(\mathbb{C})$ is a period matrix of a curve C.*

The RIEMANN relations (2.26) are necessary but not sufficient because there are much more abelian varieties than Jacobians of curves, except for the low dimensions $g = 1, 2, 3$.

Another variant of the SCHOTTKY problem is the following:
Characterize the set of points in $\mathbb{H}_g$ or in $\mathbb{H}_g/\mathbb{S}p(2g,\mathbb{Z})$ coming from Jacobians of curves via periods.

2.10. The *relative* SCHOTTKY *problem for* PICARD *curves* asks for a characterization of the period matrices of PICARD curves and the points in $\mathbb{H}_3$ or $\mathbb{H}_3/\mathbb{S}p(6,\mathbb{Z})$ coming from them.

Additionally we would like to recognize the PICARD curve C, if we know its Jacobian variety $J(C)$. Jacobians of curves have a canonical (principal) polarization coming from the intersection product of cycles. In analytic-arithmetic terms a *principal polarization* of an abelian variety $A = \mathbb{C}^g/\wedge$ is a non-degenerate skew-symmetric unimodular $\mathbb{Z}$-bilinear form $E : \wedge \times \wedge \to \mathbb{Z}$. If E is the canonical polarization of the Jacobian variety $J(C)$ of a curve C, then we set $Jac(C) = (J(C), E)$ and call this pair the *canonically polarized Jacobian (variety)* of C. We have

TORELLI's **Theorem 2.11.** *Let C, C' be smooth curves of positive genus. Then $Jac(C)$ and $Jac(C')$ are isomorphic if and only if C and C' are.*

In order to be effective we will establish and solve

2.2.3 The Effective SCHOTTKY-TORELLI Problem for PICARD Curves

Let C be a smooth PICARD curve, $\vec{\omega} = {}^t(\omega_1, \omega_2, \omega_3)$ a basis triple of $H^1 := H^0(C, \omega_C)$, $B = (\alpha_1, \ldots, \alpha_t)$ a normal basis of $H_1 := H_1(C, \mathbb{Z})$, that means $(\alpha_i \circ \alpha_j) = I_6$ (see 2.21), and

$$\Pi = \int_B \vec{\omega} = \begin{pmatrix} \boxed{\mathbf{A_1} \quad \mathbf{A_2}} & A_4 & \boxed{\mathbf{A_3}} & A_5 & A_6 \\ \bar{B}_1 & \bar{B}_2 & \bar{B}_4 & \bar{B}_3 & \bar{B}_5 & \bar{B}_6 \\ \bar{C}_1 & \bar{C}_2 & \bar{C}_4 & \bar{C}_3 & \bar{C}_5 & \bar{C}_6 \end{pmatrix} \tag{2.28}$$

the corresponding period matrix, where $\bar{}$ denotes complex conjugation.

2.12. *The* **effective** SCHOTTKY-TORELLI *problem for* PICARD *curves consists of the following program:*

1. *Find "typical" period matrices coming from* PICARD *curves and only from them.*
2. *Reconstruct typical period matrices* (2.28) *from only three entries, say from* A_1, A_2, A_3.
3. *Find a precise criterion for triples* (A_1, A_2, A_3) *to be extendable to a typical period matrix* (2.28).
4. *Given a typical period matrix* Π. *Find an effective way to establish the normal form equation* (2.1) *of a* PICARD *curve with period matrix* Π.

Remarks 2.13. The problem 1. is the relative SCHOTTKY problem in the narrow sense. The problems 2. and 3. form a bridge from 1. to 4. On the one hand they are finer parts of the first problem, on the other hand they prepare the construction of functions G_i, $i = 2, 3, 4$ depending on A_1, A_2, A_3 such that $Y^3 = X^4 + G_2X^2 + G_3X + G_4$ is the equation we look for in the fourth problem, which we understand as *effective* TORELLI *problem for* PICARD *curves.*

The restricting step 2. to a low number of matrix coefficients for characterizing typical period matrices can be roughly understood with a glance to the moduli space $M^\wedge = \mathbb{P}^2/S_4$ of PICARD curves, which is two-dimensional. It explains also that the choices of A_1, A_2, A_3 cannot be arbitrary. The relation we look for in the third problem will lead us to a symmetric domain, and we will call it the **ball criterion**.

2.3 Typical Period Matrices

Let C be a smooth curve of genus g and $G \subseteq \text{Aut } C$ a finite subgroup of the automorphism group of C. In our applications we have $g \geq 2$, hence our finiteness condition is automatically satisfied because Aut C is finite by a classical theorem.

We consider the natural representation of G in the vector space H^1 of holomorphic differential forms on C: $G \ni \gamma : \omega \mapsto \gamma^*\omega$. Locally, in a neighbourhood U of $P \in C$, ω is given by $\omega = f(t)dt$, f a holomorphic function and t a local parameter on U. On $\gamma(U)$ the differential form $\gamma^*\omega$ is represented by

$$\gamma^*\omega = \gamma^*f \cdot d\gamma^*t\,, \quad \gamma^*f = f \circ \gamma\,, \quad \gamma^*t = t \circ \gamma\,.$$

Decompose the representation into irreducible subrepresentations: $H^1 = V_1 \oplus \ldots \oplus V_\ell$. We assume that all irreducible components are one-dimensional. In our applications the group G is abelian. Then this additional condition is satisfied. Adding together isomorphic irreducible components we get a decomposition

$$H^1 = H_1^1 \oplus \ldots \oplus H_k^1\,, \quad \dim H_i^1 = g_i\,, \quad g_1 + \ldots + g_k = g \tag{2.29}$$

together with characters

$$\chi_i : G \longrightarrow \mathbb{C}^* = \mathbb{C}\backslash 0\,, \quad \gamma \longmapsto \chi_i(\gamma)\,, \; \chi_i(\gamma)\omega = \gamma^*\omega\,. \tag{2.30}$$

Definition 2.14. A basis $\vec{\omega}$ of H^1 is called *G-typical*, if it is composed of bases $\vec{\Omega}_i$ of H_i^1:

$$\vec{\omega} = {}^t(\vec{\omega}_1, \ldots, \vec{\omega}_g) = {}^t(\vec{\Omega}_1, \ldots, \vec{\Omega}_k) \tag{2.31}$$

Example 2.15. A *cycloelliptic (cyclic, superelliptic) curve* is a plane projective curve of equation type $C : Y^m = p_n(X)$, p_n a polynomial of degree n. Then we have an obvious action of the cyclic group $G = \mathbb{Z}/m\mathbb{Z} = \langle\gamma\rangle$, say, on C defined by $\gamma : C \xrightarrow{\sim} C$, $(x, y) \mapsto (x, \zeta_m y)$, ζ_m a primitive m-th unit root. Now let $C : Y^3 = p_4(X)$ be a smooth PICARD curve. Then $G = \mathbb{Z}/3\mathbb{Z}$. The G-typical basis

$$\vec{\omega} = {}^t\left(dx/y, dx/y^2, xdx/y^2\right) \tag{2.32}$$

is called the *typical H^1-basis* of C. Setting $\delta^2 = \zeta_3$ we find with the above notations

$$\begin{aligned} \gamma^*(dx/y) &= \gamma^*(1/y)d\gamma^*x = (y/\delta^2)dx = \delta dx/y\,, \\ \gamma^*(dx/y^2) &= (y^2/\delta)dx = \delta^2 dx/y\,,\ \gamma^*(xdx/y^2) = \delta^2 xdx/y^2\,. \end{aligned} \tag{2.33}$$

We set

$$H^1 = H_1^1 \oplus H_2^1 = H_\delta^1 \oplus H_{\bar{\delta}}^1 \quad \text{and} \quad \vec{\omega} = (\omega_1, \vec{\Omega}_1)\,, \tag{2.34}$$

where

$$\omega_1 = dx/y\,, \quad \vec{\Omega}_2 = {}^t(dx/y^2, xdx/y^2)\,.$$

2.3.1 G-typical Bases of $H_1(\mathbb{Z})$

We come back to the general situation described at the begin of this section. Set, as before, $H_1 = H_1(\mathbb{Z}) = H_1(C, \mathbb{Z})$. Let

$$\mathbb{Z}[G] = \left\{ \sum_{\gamma \in G} m_\gamma \gamma ;\ m_\gamma \in \mathbb{Z} \right\}$$

be the group ring of G over $\mathbb{Z}$. Via the action

$$G \times H_1 \longrightarrow H_1 , \quad (\gamma, \alpha) \longmapsto \gamma \circ \alpha$$

the abelian group H_1 is a $\mathbb{Z}\ [G]$-module. Let $2g = rkH_1 = rs$ for two natural numbers r, s. The product of group rings $\mathbb{Z}[G]^r$ acts componentwise on H_1^r:

$$(Z[G]^r \times H_1^r) \ni (\Delta, A) \longmapsto \Delta A \in H_1^r$$

This way we get a pairing

$$\begin{aligned} * : (\mathbb{Z}[G]^r)^s \times H_1^r &\longrightarrow H_1^{2g} , \\ (\mathbb{G}, A) = ((\mathbb{G}_1; \ldots; \mathbb{G}_s), A) &\longmapsto (\mathbb{G}_1 A; \ldots; \mathbb{G}_s A) = \mathbb{G} * A \end{aligned}$$

with obvious notation. Moreover, the symmetric group S_{2g} acts from the right-hand side on H_1^{2g} by permutation of components. Composing with the above pairing we set

$$\mathbb{G} *_S A = (\mathbb{G} * A)S = (\mathbb{G}_1 A; \ldots; \mathbb{G}_s A) \cdot S,\ S \in S_{2g}. \tag{2.35}$$

Definition 2.16. A basis B of $H_1(C, \mathbb{Z})$ is called G-*typical* (of decomposition) type $(r, s, S))$, if B is a normal basis and $B = \mathbb{G} *_S B_0$ for suitable $\mathbb{G} \in (Z[G]^r)^s$, $B_0 \in H_1^r$, $S \in S_{2g}$. More precisely, we will call it also a $\mathbb{G}$-*typical basis* (*of type* S).

Proposition 2.17. *On each smooth* Picard *curve* $C : Y^3 = p_4(X)$ *exists a* $\mathbb{G}$-*typical basis of type* $(3, 4) \in S_6$ *with* $\mathbb{G} = (1, 1, 1; -\gamma, \gamma, \gamma^2)$, $\langle\gamma\rangle = G = \mathbb{Z}/3\mathbb{Z}$ *as in* 2.15.

Proof: The quotient map $C \to C/G = \mathbb{P}^1$ is a three-sheeted covering of $\mathbb{P}^1$ with five branch points $\infty, a_1, a_2, a_3, a_4$ (the zeros of $p_4(X)$). The picture (2.36) represents C as union of three copies of the Gauss plane, where we identify the branch points marked in each sheet lying over each other. The four arising ramification points are also denoted by $a_1, \ldots, a_4$. Keep in mind that the same happens at infinity. Along four cuts joining a_i with ∞ in each sheet, the copies are joint with each other. Crossing the cuts we come from one sheet to the other as described in picture (2.36), where we draw a loop (cycle) representing an element of $H_1(\mathbb{Z})$.

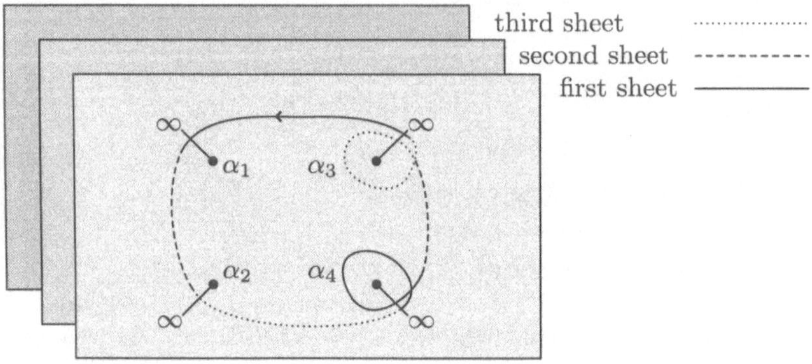

(2.36)

The automorphism $\gamma : C \to C$, $(x, y) \mapsto (x, \delta^2 y)$ acts by changing the sheets. Drawing only four cuts and loops crossing them we define in picture (2.37) six cycles $\alpha_i, i = 1, \ldots, 6$. Obviously it holds that $(\beta_i \circ \beta_j) = I_6$, where

$$B = (\beta_1, \ldots, \beta_6) = (\alpha_1, \ldots, \alpha_6)(3, 4) .$$

Therefore B defines a normal basis of $H_1(C, \mathbb{Z})$. Preserving notations

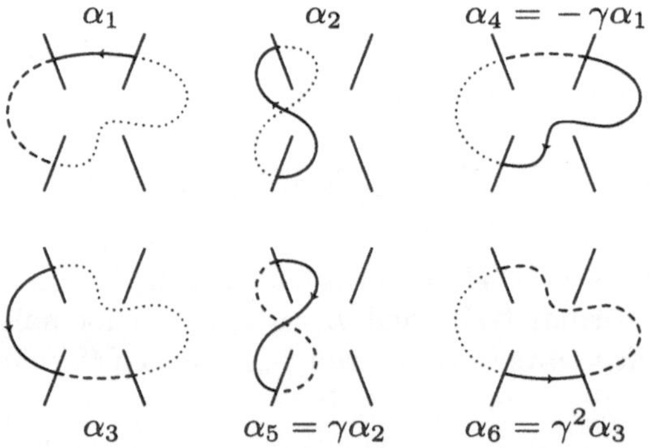

(2.37)

during the change to cycle classes we see additionally that

$$B = \mathbb{G} *_{(3,4)} B_0 = (\alpha_1, \alpha_2, \alpha_3; -\gamma\alpha_1, \gamma\alpha_2, \gamma^2\alpha_3)(3, 4) \tag{2.37}$$

with

$$B_0 = (\alpha_1, \alpha_2, \alpha_3) .$$

Therefore B is a $\mathbb{G}$-typical H_1-basis on C of type (3,4). The proposition is proved. □

Definition 2.18. A G-typical H_1-basis of permutation type (3,4) on a PICARD curve C with the fixed 6-tuple G defined in 2.17 is called shortly a *typical H_1-basis* on C. We will also call B_0 a *typical $\mathcal{O}_K$-basis*, $\mathcal{O}_K$ the ring of EISENSTEIN numbers (see section 2.5), if $B = \mathbb{G} *_{(3,4)} B_0$ is typical.

2.3.2 Period Matrices of G-typical Bases

We go back to the general situation described at the begin of this section. We have G-typical H_1-bases (2.31) and on G-typical H_1-bases defined in 2.16:

$$\vec{\omega} = {}^t(\vec{\Omega}_1, \ldots, \vec{\Omega}_k)\ , \quad B = \mathbb{G} *_S B_0 = (\mathbb{G}_1; \ldots ; \mathbb{G}_s) *_S B_0 \tag{2.39}$$

Let $\chi : G \to \mathbb{C}^\times$ be a character. It extends linearly to

$$\chi : \mathbb{Z}[G] \longrightarrow \mathbb{C}\ .$$

We define

$$\chi(\mathbb{G}) = (\chi(\mathbb{G}_1), \ldots, \chi(\mathbb{G}_S)) = (\chi(\gamma_1), \ldots, \chi(\gamma_g))\ ,$$

$$\vec{a} *_S {}^t\vec{z} = {}^t(\vec{a}z_1, \ldots, \vec{a}z_m)S, \qquad \vec{a} \in \mathbb{C}^n,\ \vec{z} \in \mathbb{C}^m,\ S \in S_n\ .$$

Lemma 2.19. *With the above notations the period matrix corresponding to $\vec{\omega}$, B looks like*

$$\Pi(\vec{\omega}, B) = \begin{pmatrix} \chi_1(\mathbb{G}) *_S \int_{B_0} \vec{\Omega}_1 \\ \cdots\cdots\cdots \\ \chi_k(\mathbb{G}) *_S \int_{B_0} \vec{\Omega}_k \end{pmatrix}. \tag{2.40}$$

Proof: This follows immediately from the substitution rule $\int_{\gamma\alpha} \omega = \int_\alpha \gamma^*\omega$ for $\alpha \in H_1$, $\omega \in H^1$, hence

$$\int_{\gamma\alpha} \omega = \int_\alpha \gamma^*\omega = \int_\alpha \chi_i(\gamma)\omega = \chi_i(\gamma) \int_\alpha \omega$$

for $\gamma \in G$, $\omega \in H^1_i$. Therefore

$$\begin{aligned}\int_B \vec{\Omega}_i &= \int_{\mathbb{G} *_S B_0} \vec{\Omega}_i \\ &= \left(\int_{\mathbb{G}_1 B_0} \vec{\Omega}_i, \ldots, \int_{\mathbb{G}_s B_0} \vec{\Omega}_i \right) \cdot S \\ &= \left(\chi_i(\mathbb{G}_1) \int_{B_0} \vec{\Omega}_i, \ldots, \chi_i(\mathbb{G}_S) \int_{B_0} \vec{\Omega}_1 \right) \cdot S \\ &= \chi_i(\mathbb{G}) *_S \int_{B_0} \vec{\Omega}_i\ ,\end{aligned}$$

hence

$$\Pi(\vec{\omega}, B) = \int_B \vec{\omega} = \begin{pmatrix} \int_B \vec{\Omega}_1 \\ \cdots\cdots \\ \int_B \vec{\Omega}_k \end{pmatrix} = \begin{pmatrix} \chi_1(\mathbb{G}) *_S \int_{B_0} \vec{\Omega}_1 \\ \cdots\cdots\cdots\cdots \\ \chi_k(\mathbb{G}) *_S \int_{B_0} \vec{\Omega}_k \end{pmatrix}.$$

□

Definition 2.20. Period matrices of type (2.40) are called *G-typical* (or *$\mathbb{G}$-typical*). In the case of PICARD curves with $S = (3, 4)$, $\mathbb{G} = (1, 1, 1; -\gamma, \gamma, \gamma^2)$ we call them simply *typical period matrices.*

With the notations of (2.34) the typical period matrices of PICARD curves look like

$$\int_B \vec{\omega} = \int_B \begin{pmatrix} \omega_1 \\ \vec{\Omega}_2 \end{pmatrix} = \begin{pmatrix} \chi_1(\mathbb{G}_1) \int_{B_0} \omega_1; \chi_1(\mathbb{G}_2) \int_{B_0} \omega_1 \\ \chi_2(\mathbb{G}_1) \int_{B_0} \vec{\Omega}_2; \chi_2(\mathbb{G}_2) \int_{B_0} \vec{\Omega}_2 \end{pmatrix} (3, 4).$$

Now we set

$$\int_{B_0} \omega_1 = (A_1, A_2, A_3) \,, \quad \int_{B_0} \vec{\Omega}_2 = \begin{pmatrix} \bar{B}_1, \bar{B}_2, \bar{B}_3 \\ \bar{C}_1, \bar{C}_2, \bar{C}_3 \end{pmatrix}. \tag{2.41}$$

Lemma 2.21. *With the notations of* (2.20) *each typical period matrix of a smooth* PICARD *curve can be written as*

$$\Pi = \begin{pmatrix} A_1, A_2, A_3; & -\delta A_1, \delta A_2, \bar{\delta} A_3 \\ \bar{B}_1, \bar{B}_2, \bar{B}_3; & -\bar{\delta} \bar{B}_1, \bar{\delta} \bar{B}_2, \delta \bar{B}_3 \\ \bar{C}_1, \bar{C}_2, \bar{C}_3; & -\bar{\delta} \bar{C}_1, \bar{\delta} \bar{C}_2, \delta \bar{C}_3 \end{pmatrix} \cdot (3, 4).$$

□

2.4 Metrization

Remember that $I = I_6 = \left(\begin{array}{c|c} 0 & E_3 \\ \hline -E_3 & 0 \end{array}\right)$. With the hermitian matrix $J = I/\sqrt{-3} = \frac{-i}{\sqrt{3}} I$ we define the hermitian product $\langle \ , \ \rangle$ on $\mathbb{C}^6$ by $\langle \mathcal{A}, \mathcal{B} \rangle = \mathcal{A} J^t \bar{\mathcal{B}}$.

Now we fix $\omega = \omega_1 = dx/y$ on a smooth PICARD curve $C : Y^3 = p_4(X)$. Looking at the first row of the typical period matrix (2.41) we define the linear embedding

$$* : \mathbb{C}^3 \hookrightarrow \mathbb{C}^6 \tag{2.42}$$
$$\mathcal{A} = (A_1, A_2, A_3) \mapsto *\mathcal{A} = (A_1, A_2, -\delta A_1, A_3, \delta A_2, \bar{\delta} A_3).$$

Without change of notations we extend the map $\int \omega : H_1 \to \mathbb{C}$ to powers of H_1 to obtain the following commutative diagram:

$$\begin{array}{ccccccc} \{\text{typical } O\text{-bases on } C\} & \overset{*}{\underset{*(3,4)}{\hookrightarrow}} & \{\text{normal } \mathcal{O}\text{-bases on } C\} & & B_0 & \mapsto & *B_0 \\ \downarrow & & \downarrow & & \downarrow & & \downarrow \\ \mathbb{C}^3 & \hookrightarrow & \mathbb{C}^6 & & \int\limits_{B_0}\omega & \mapsto & \int\limits_{*B_0}\omega \end{array} \qquad (2.43)$$

Keep in mind that we write from now on $*$ instead of $*_{(3,4)}$. Along $*$ we restrict the hermitean bilinear form $\langle\ ,\ \rangle$ on $\mathbb{C}^6$ to $\mathbb{C}^3$:

$$\langle\ ,\ \rangle : \mathbb{C}^3 \times \mathbb{C}^3 \to \mathbb{C}\,, \quad \langle \mathfrak{a}, \mathfrak{b}\rangle =_{Df} \langle *\mathfrak{a}, *\mathfrak{b}\rangle = *\mathfrak{a} J^t \overline{*\mathfrak{b}} \qquad (2.44)$$

Next we fix primitive third unit roots setting

$$\rho = \delta = e^{2\pi i/3}. \qquad (2.45)$$

The orthogonal relation with respect to $\langle\ ,\ \rangle$ is denoted by $\perp$.

Lemma 2.22. *Let* $\Pi = \begin{pmatrix} *\mathfrak{a} \\ \overline{*\mathfrak{b}} \\ \overline{*\mathfrak{e}} \end{pmatrix}$ *be a typical period matrix. Then it holds that*

$$\text{(i) } \langle \mathfrak{a}, \mathfrak{a}\rangle < 0\,, \quad \text{(ii) } \mathfrak{a} \perp \mathfrak{b}, \mathfrak{e}\,, \quad \text{(iii) } \mathfrak{b}, \mathfrak{e} \text{ are linearly independent.} \qquad (2.46)$$

Proof: Remember the RIEMANN period relations (2.26)

$$\Pi J^t \Pi = 0\,, \quad \Pi J^t \bar{\Pi} < 0\,.$$

(ii) $0 = \overline{*\mathfrak{b}} J^t(*\mathfrak{a}) = \overline{*\mathfrak{e}} J^t(*\mathfrak{a})$ implies $\langle \mathfrak{a}, \mathfrak{b}\rangle = \langle \mathfrak{a}, \mathfrak{e}\rangle = 0$.

(i) $\langle \mathfrak{a}, \mathfrak{a}\rangle = (*\mathfrak{a}) J^t(\overline{*\mathfrak{a}}) = (1,0,0)\Pi J^t \bar{\Pi} \begin{pmatrix} 1 \\ 0 \\ 0 \end{pmatrix} < 0.$

(iii) follows from $rk(\Pi) = 3$.

Now we determine the signature of $\langle , \rangle$ by direct computations. We work with $\mathfrak{a} = (A_1, A_2, A_3)$, $\mathfrak{b} = (B_1, B_1, B_3) \in \mathbb{C}^3$. Then we find

$$
\begin{aligned}
&\langle \mathfrak{a}, \mathfrak{b} \rangle \\
&= (*\mathfrak{a}) J^t (*\mathfrak{b}) \\
&= \frac{-i}{\sqrt{3}} (A_1, A_2, -\delta A_1, A_3, \delta A_2, \bar{\delta} A_3,) \left(\begin{array}{ccc|ccc} & \bigcirc & & 1 & 0 & 0 \\ & & & 0 & 1 & 0 \\ & & & 0 & 0 & 1 \\ \hline -1 & 0 & 0 & & & \\ 0 & -1 & 0 & & \bigcirc & \\ 0 & 0 & -1 & & & \end{array} \right) \begin{pmatrix} \bar{B}_1 \\ \bar{B}_2 \\ -\bar{\delta} \bar{B}_1 \\ \bar{B}_3 \\ \bar{\delta} \bar{B}_2 \\ \delta \bar{B}_3 \end{pmatrix} \\
&= \frac{-i}{\sqrt{3}} (-A_3 \bar{B}_1 - \delta A_1 \bar{B}_2 + \delta A_3 \bar{B}_1 + A_1 \bar{B}_3 + \bar{\delta} A_2 \bar{B}_2 - \bar{\delta} A_1 \bar{B}_3) \\
&= \frac{-i}{\sqrt{3}} (A_3 \bar{B}_1 (\delta - 1) + A_2 \bar{B}_2 (\bar{\delta} - \delta) + A_1 \bar{B}_3 (1 - \bar{\delta})) \\
&= \bar{\delta} A_3 \bar{B}_1 + A_2 \bar{B}_2 + \delta A_1 \bar{B}_3 = \mathfrak{a} \begin{pmatrix} 0 & 0 & \delta \\ 0 & 1 & 0 \\ \bar{\delta} & 0 & 0 \end{pmatrix} {}^t \bar{\mathfrak{b}} ,
\end{aligned}
$$

hence

$$
\langle \mathfrak{a}, \mathfrak{b} \rangle = \mathfrak{a} \begin{pmatrix} 0 & 0 & \rho \\ 0 & 1 & 0 \\ \bar{\rho} & 0 & 0 \end{pmatrix} {}^t \bar{\mathfrak{b}} , \quad \mathfrak{a}, \mathfrak{b} \in \mathbb{C}^3 , \ \rho = e^{2\pi i/3} . \tag{2.47}
$$

With

$$
M = \begin{pmatrix} \rho & 0 & -1 \\ 0 & 1 & 0 \\ -\bar{\rho} & 0 & -1 \end{pmatrix} \in \mathbb{G}l_3(\mathcal{O}), \quad \mathcal{O} = \mathcal{O}_K, \quad K = \mathbb{Q}(\sqrt{-3})
$$

we see that

$$
M \begin{pmatrix} 1 & 0 & 0 \\ 0 & 1 & 0 \\ 0 & 0 & -1 \end{pmatrix} {}^t \bar{M} = \begin{pmatrix} 0 & 0 & \rho \\ 0 & 1 & 0 \\ \bar{\rho} & 0 & 0 \end{pmatrix} .
$$

Consequently, we obtain the following

Lemma 2.23. *The signature of* $\langle , \rangle$ *on* $\mathbb{C}^3$ *is (2,1).*

The hermitian bilinear form $\langle , \rangle$ above defines a two-ball in $\mathbb{P}^2$, namely

$$
\mathbb{B} = \left\{ \mathbb{P}\mathfrak{a} = (A_1 : A_2 : A_3) ; \ \mathfrak{a} = (A_1, A_2, A_3) \in \mathbb{C}^3, \langle \mathfrak{a}, \mathfrak{a} \rangle < 0 \right\} . \tag{2.48}
$$

From the above transformations it is clear that $\mathbb{B}$ is a $\mathbb{G}l_3(K)$-shift of the standard two-ball

$$
\mathbb{B}^2 = \left\{ (z_1, z_2) \in \mathbb{C} ; \ |z_1|^2 + |z_2|^2 < 1 \right\}
$$

in $\mathbb{P}^2$, namely $\mathbb{B}\mathbb{P}(M) = \mathbb{B}^2$, K as below.

2.5 Arithmetization

The imaginary quadratic field $K = \mathbb{Q}(\sqrt{-3})$ is called the field of EISENSTEIN *numbers.* Its ring of integers

$$\mathcal{O} = \mathcal{O}_K = \left\{a/2 + b\sqrt{-3}/2 \,;\; a, b \in \mathbb{Z}\,,\; a \equiv b \bmod 2\right\} = \mathbb{Z} + \mathbb{Z}\rho$$

is called the ring of EISENSTEIN *integers.*

Now let $C : Y^3 = p_4(X)$ be a smooth PICARD curve.

2.24. $H_1(C, \mathbb{Z})$ is a free $\mathcal{O}$-module of $\mathcal{O}$-rank 3.

Proof: If $\alpha : [0,1] \to C$, $t \mapsto (x(\alpha(t)), y(\alpha(t)))$ is a cycle on C, then $\rho\alpha$ is defined to be the cycle $t \mapsto (x(\alpha(t)), \rho y(\alpha(t)))$. So we obtain an action of $\mathcal{O}$ on $H_1(\mathbb{Z})$. If $a\beta = 0$ for $a \in \mathcal{O}$, $\beta \in H_1$, then also $\bar{a}a\beta = 0$, hence $\beta = 0$ or $a\bar{a} = 0$, that means $a = 0$, because H_1 has no $\mathbb{Z}$-torsion. Moreover, $\mathcal{O}$ is a principal domain. Therefore H_1 is a free $\mathcal{O}$-module. Comparing $\mathbb{Z}$-ranks we recognize that

$$H_1(C, \mathbb{Z}) \underset{\mathcal{O}}{\cong} \mathcal{O}^3.$$

□

In (2.22) we defined the symplectic modular group $\mathbb{S}p(2g, \mathbb{Z})$, and we saw how it acts on the space of period matrices of a curve of genus g, see (2.23). Especially, if C is a PICARD curve, then $\mathbb{S}p(6, \mathbb{Z})$ acts from the right on the space of period matrices of C coming from the action on normal bases. This is not longer true if we restrict ourselves to typical bases or typical period matrices.

Notation 2.25. We denote by $\mathbb{S}p'(6, \mathbb{Z})$ the subgroup of $\mathbb{S}p(6, \mathbb{Z})$ of all elements transferring a typical basis to a typical basis on a smooth PICARD curve C.

We will see that the definition does depend neither on the special choice of the PICARD curve nor on the choice of a typical basis on it.

First we introduce the *unitary group*

$$\mathbb{U}((2,1), \mathbb{C}) = \left\{g \in \mathbb{G}l_3(\mathbb{C})\,;\; \langle g\mathfrak{a}, g\mathfrak{b}\rangle = \langle \mathfrak{a}, \mathfrak{b}\rangle \text{ for all } \mathfrak{a}, \mathfrak{b} \in \mathbb{C}^3\right\}.$$

It acts on the ball $\mathbb{B}$ in obvious manner:

$$\mathbb{U}((2,1), \mathbb{C}) \ni g : \mathbb{B} \ni \beta = \mathbb{P}\mathfrak{b} \mapsto \mathbb{P}(\mathfrak{b}g) \in \mathbb{B}$$

Definition 2.26. The *(full)* PICARD *modular group* (of EISENSTEIN *numbers)* is the arithmetic subgroup

$$\mathbb{U}((2,1), \mathcal{O}) = \mathbb{G}l_3(\mathcal{O}) \cap \mathbb{U}((2,1), \mathbb{C})$$

of $\mathbb{U}((2,1), \mathbb{C})$.

Lemma 2.27. *The groups* $\mathbb{S}p'(6,\mathbb{Z})$ *and* $\mathbb{U}((2,1),\mathcal{O})$ *are isomorphic.*

Proof: We choose a typical $\mathcal{O}$-basis B_0 of $H_1 \cong \mathcal{O}^3$, $G' \in \mathbb{S}p'(6,\mathbb{Z})$ and define $g' \in \mathbb{G}l_3(\mathcal{O})$ by $B_0\, g' = B_0'$ with $*B_0' = (*B_0)G'$ a typical $\mathbb{Z}$-basis and B_0' a typical $\mathcal{O}$-basis of H_1, by assumption. So we have a correspondence $G' \mapsto g' \in \mathbb{G}l_3(\mathcal{O})$. Along the vertical maps of (2.43) we transfer g' and G' to actions from the right on $\mathbb{C}^3$ respectively on $\mathbb{C}^6$. For $\mathfrak{a},\mathfrak{b} \in \mathbb{C}^3$ we get with obvious notations

$$\begin{aligned}\langle \mathfrak{a}g, \mathfrak{b}g\rangle &= \langle *(\mathfrak{a}g), *(\mathfrak{b}g)\rangle \\ &= \langle (*\mathfrak{a})G, (*\mathfrak{b})G\rangle \\ &= (*\mathfrak{a})GJ^tG^t\overline{(*\mathfrak{b})} \\ &= (*\mathfrak{a})J^t\overline{(*\mathfrak{b})} \\ &= \langle *\mathfrak{a}, *\mathfrak{b}\rangle \\ &= \langle \mathfrak{a}, \mathfrak{b}\rangle\end{aligned}$$

because G is symplectic. Thus $g \in \mathbb{U}((2,1),\mathbb{C}) \cap \mathbb{G}l_3(\mathcal{O}) = \mathbb{U}((2,1),\mathcal{O})$. We can extend the actions to period matrices, especially to typical period matrices Π described in 2.21, 2.22. It is not difficult to see that g and G correspond to each other by the identity

$$\begin{pmatrix} *(\mathfrak{a}g) \\ \overline{*(\mathfrak{b}g)} \\ \overline{*(\mathfrak{e}\,g)} \end{pmatrix} = \begin{pmatrix} *\mathfrak{a} \\ \overline{*\mathfrak{b}} \\ \overline{*\mathfrak{e}} \end{pmatrix} G\,. \tag{2.50}$$

□

Corollary 2.28. *Let* A *be a typical* $\mathcal{O}$*-basis of a smooth* PICARD *curve* C. *Then* A' *is a typical* $\mathcal{O}$-basis *iff* A' *belongs to the* $\mathbb{U}((2,1),\mathcal{O})$-orbit *of* A.

Definition 2.29. For a smooth PICARD curve C and a typical $\mathcal{O}$-basis A of $H_1(C,\mathbb{Z})$ we call $\mathfrak{a} = \int_A dx/y \in \mathbb{C}^3$ a *C-typical vector.* Its projection $\mathbb{P}\mathfrak{a} \in \mathbb{B} \subset \mathbb{P}^2$ (see (2.46), Lemma 2.23) is called a *C-typical ball point.*

Corollary 2.30. *Let* $\mathfrak{a}$ *be a* C*-typical vector. Then* $\mathfrak{b}$ $(\mathbb{P}\mathfrak{b})$ *is a* C*-typical vector (ball point) iff it belongs to the* $\mathbb{U}((2,1),\mathcal{O})$*-orbit of* $\mathfrak{a}$ $(\mathbb{P}\mathfrak{a})$.

We see that each smooth PICARD curve C defines via typical period matrices an infinite discrete set of C-typical points in the ball $\mathbb{B}$. If we move continuously our starting curve C in a family of PICARD curves, then the corresponding typical points move continuously in the ball $\mathbb{B}$. The question is, which part of $\mathbb{B}$ is filled by typical points of (smooth) PICARD curves? Before we answer it we put in the following section.

2.6 A Retrospect to Elliptic Curves

On the POINCARÉ upper half plane $\mathbb{H}$ the group $\mathbb{S}l_2(\mathbb{R})$ acts by fractional linear transformation (see 1.1). We remark that it is easy to find a fractional linear transformation in $\mathbb{G}l_2(\mathbb{C})$ mapping $\mathbb{H}$ biholomorphically onto the unit disc $\mathbb{D} = \mathbb{B}^1 = \{z \in \mathbb{C}; |z| < 1\}$. The group of biholomorphic automorphisms of $\mathbb{D}$ is $\mathbb{PU}((1,1), \mathbb{C})$.

There is an arithmetic connection between $\mathbb{H}$ and $\mathbb{P}^1 = \mathbb{P}^1(\mathbb{C})$ arranged by the analytic family $\mathcal{E}/\mathbb{H}$, the modular group $\mathbb{S}l_2(\mathbb{Z})$ and the very explicit elliptic modular function j described in the following scheme (2.51), see (1.1):

$$\begin{array}{ll} \mathbb{H} & \mathcal{E} = \{E_\tau\}_{\tau\in\mathbb{H}}\,,\; E_\tau : Y^2 = 4X^3 - g_2(\tau)X - g_3(\tau) \\ j \downarrow \mathbb{S}l_2(\mathbb{Z}) & \\ \mathbb{P}^1 \setminus \{\infty\} & \text{Moduli space: } E_\tau \cong E_{\tau'} \Longleftrightarrow \tau' \in \mathbb{S}l_2(\mathbb{Z}) \cdot \tau \end{array} \tag{2.51}$$

We are inspired to find a similar scheme (2.52) with two-dimensional base spaces. The contributions should be described as explicit as possible.

$$\begin{array}{ll} \mathbb{B} & \mathcal{C} = \{C_\tau\}_{\tau\in\mathbb{B}}\,,\; C_\tau : Y^3 = 4X^4 + G_2(\tau)X^2 + G_3(\tau)X + G_4(\tau) \\ J \downarrow \mathbb{G} & \\ \mathbf{M} & \text{moduli space: } C_\tau \cong C_{\tau'} \Longleftrightarrow \tau \in \Gamma\tau' \end{array} \tag{2.52}$$

The moduli space $\mathbf{M} = \left(\mathbb{P}^2 \setminus \{P_1, \ldots, P_4\}\right)/S_4$ has been found in (2.1). The group Γ should be an arithmetic subgroup of $\mathbb{U}((2,1,\mathbb{C})$. By our arithmetization in the last section a natural candidate is the PICARD modular group $\mathbb{U}((2,1), \mathcal{O})$. Suitable holomorphic functions $G_i(\tau)$ and the holomorphic uniformization map J have to be constructed.

What we found until now is a multivalued inverse map

$$\begin{array}{rcl} \mathbf{M} = (\mathbb{P}^2 \setminus \triangle\!\!\!\!\star)/S_4 & \dashrightarrow & \mathbb{B}\,, \\ cl(C) & \longmapsto & \{C\text{-typical points}\} \end{array} \tag{2.53}$$

We have an analogous situation in the theory of elliptic curves. The inverse of the uniformizing map j in (2.51) comes out from the theory of elliptic integrals. Consider the elliptic integrals (of first kind)

$$z(s) = \int_{s_0}^{s} dx/\sqrt{x(x-1)(x-t)}\,, \quad t \in \mathbb{C} \setminus \{0,1\}\,.$$

The integrand can be understood as the holomorphic differential form $\omega = dx/y$ on the elliptic curve $E_t : Y^2 = X(X-1)(X-t)$ and the integral can be taken along paths on E_t joining points s_0 and s on E. Since E_t is not simply-connected, the value $z(s)$ depends on the choice of paths. But it is unique modulo the lattice $\wedge_t$ of period integrals $\int_\alpha \omega$, $\alpha \in H_1(E_t, \mathbb{Z})$.

ABEL and JACOBI studied the inverse function $s(z)$. It is a meromorphic $\wedge_t$-periodic function on $\mathbb{C}$, hence an elliptic function. So the study of elliptic integrals is closely connected with the investigation of period integrals on elliptic curves.

According to their dependence on t the period integrals

$$\eta(t) = \int_{\alpha_t} dx/\sqrt{x(x-1)(x-t)}\ , \quad \alpha_t \text{ cycle on } E_t$$

are (special) *hypergeometric integrals.* The hypergeometric function η is multivalued because it depends on the choice of cycles. In order to understand the multivalence one considers in a small simply-connected neighbourhood of t_0, say, two independent cycle families $\alpha_1(t)$, $\alpha_2(t)$. The corresponding quotient function $I(t) = \eta_1(t)/\eta_2(t)$ is holomorphic around t_0. After analytic extension it becomes multivalued again. The values lie in the disc $\mathbb{D}$.

More precisely, we consider two cuts along the real line in $\mathbb{C}$ joining $0, 1$ or $1, \infty$, respectively. If we move t in $\mathbb{C}\backslash\{0,1\}$, then the values $I(t)$ move in $\mathbb{D}$. There is a nice (infinite non-euclidean) triangulation of $\mathbb{D}$ drawn in picture (2.54). If t crosses one of our

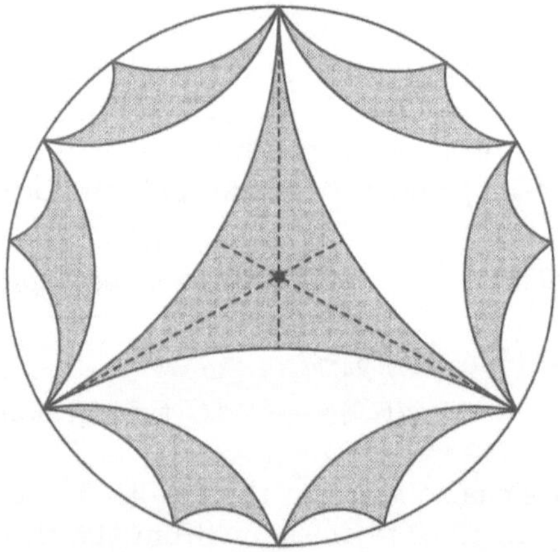

(2.54)

cuts, then $I(t)$ crosses the boundary of one of the triangles in the picture (2.54).

There is a biholomorphic map from the disc onto the upper half plane $\mathbb{H}$. It transfers the action of $\mathbb{S}l_2(\mathbb{R})$ on $\mathbb{H}$ to an action on $\mathbb{D}$ by conjugation with an element of $\mathbb{G}l_2(\mathbb{C})$. Especially, $\mathbb{S}l_2(\mathbb{Z})$ acts on $\mathbb{D}$ in this way.

The big triangles with tree cusps on the boundary circle of the disc are fundamental domains of the congruence subgroup $\Gamma(2)$ of $\mathbb{S}l_2(\mathbb{Z})$ defined by the exact sequence

$$1 \longrightarrow \Gamma(2) \longrightarrow \mathbb{S}l_2(\mathbb{Z}) \longrightarrow \mathbb{S}l_2(\mathbb{Z}/2\mathbb{Z}) \longrightarrow 1\ . \qquad (2.55)$$

The smaller triangles in (2.54) with partly dotted boundary appear as fundamental domains of the full modular group $\mathbb{S}l_2(\mathbb{Z})$. The infinite triangulation (2.54) corresponds to the decomposition (1.3) of the upper half plane $\mathbb{H}$. The factor group $\mathbb{S}l_2(\mathbb{Z})/\Gamma(2)$ is isomorphic to the symmetric group S_3. We get a commutative diagram

$$
\begin{array}{ccc}
& \mathbb{H} & \\
& \Big\downarrow{\scriptstyle\Gamma(2)} \quad \searrow{\scriptstyle\mathbb{S}l_2(\mathbb{Z})} & \\
& & j \\
\mathbb{C}\backslash\{0,1\} \longrightarrow & (\mathbb{P}^1\backslash\{0,1,\infty\}) \xrightarrow{\ S_3\ } & (\mathbb{P}^1\backslash\{\infty\}) \, .
\end{array}
\tag{2.56}
$$

We change to the next higher dimension 2, and remove in $\mathbb{C}^2$ the following five lines:

$$
\mathbb{C}^2 \supset \boxslash \ : t_1, t_2 = 0, 1, \ t_1 = t_2 \tag{2.56}
$$

Around $t_0 \in \mathbb{C}^2\backslash\boxslash$ we choose sufficiently independent cycles $\alpha_1(t), \alpha_2(t), \alpha_3(t) \in H_1(C_t, \mathbb{Z})$ depending continuously on t. Then we set

$$
I_i(t) = \int_{\alpha_i(t)} dx/y, \ i = 1, 2, 3 \, ; \quad I(t) = (I_1(t) : I_2(t) : I_3(t)) \, . \tag{2.58}
$$

Now we assume especially that the $\alpha_i(t)$ form a typical $\mathcal{O}$-basis in the sense of Definition 2.18. Then we know that $I(t)$ lies in the ball $\mathbb{B}$, see 2.29. In analogy to (2.56) we would like to get a commutative diagram

$$
\begin{array}{ccc}
& \mathbb{B} \subset \mathbb{P}^2 & \\
& \Big\downarrow{\scriptstyle\Gamma'} \quad \searrow{\scriptstyle\Gamma} & \\
t \in \mathbb{C}\backslash\boxslash \longrightarrow & (\mathbb{P}^2\backslash\{\text{points}\}) \xrightarrow{\ S_4\ } & (\mathbb{P}^2\backslash\{\text{points}\})/S_4
\end{array}
\tag{2.59}
$$

via analytic extension. This would give us a nice analytic understanding of the correspondence (2.53), see also (2.52). We see that the PICARD family (of ordered PICARD curves) (2.15) plays an important role.

2.7 Rough Solution of the Relative SCHOTTKY Problem for PICARD Curves

With the notations of Lemma 2.22 we introduce

$$
\tilde{\mathbb{B}} = \left\{ \Pi(\mathfrak{a}, \mathfrak{b}, \mathfrak{e}) := \begin{pmatrix} *\mathfrak{a} \\ \overline{*\mathfrak{b}} \\ \overline{*\mathfrak{e}} \end{pmatrix} ; \ \langle \mathfrak{a}, \mathfrak{b} \rangle < 0, \ \mathfrak{b}, \mathfrak{e} \text{ basis of } \mathfrak{a}^\perp \right\}
$$

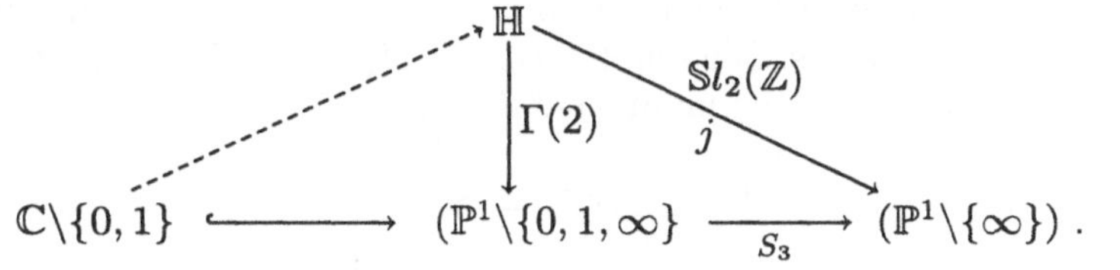

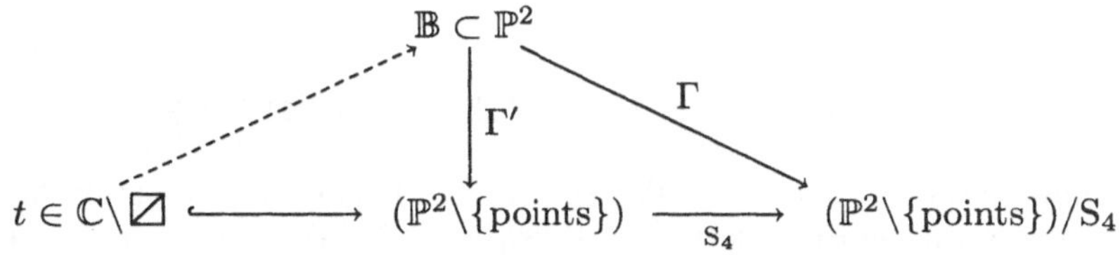

extending the analytic space

$$\tilde{\mathbb{B}}_0 = \{\text{typical period matrices of smooth PICARD curves}\}.$$

The diagonally embedded subgroup $\mathbb{G}l_1 \times \mathbb{G}l_2$ of $\mathbb{G}l_3 = \mathbb{G}l_3(\mathbb{C})$ acts from the left-hand side on $\tilde{\mathbb{B}}$. The restricted action on $\tilde{\mathbb{B}}_0$ corresponds to changes of typical H^1-bases on PICARD curves. Sending the coset $(\mathbb{G}l_1 \times \mathbb{G}l_2)\Pi(\mathfrak{a}, \mathfrak{b}, \mathfrak{e})$ to $\mathbb{P}\mathfrak{a}$ we obtain a biholomorphic map and a commutative diagram

$$\begin{array}{ccc} \mathbb{G}l_1 \times \mathbb{G}l_2\backslash\tilde{\mathbb{B}} & \xrightarrow{\sim} & \mathbb{B} \\ \uparrow & & \uparrow \\ \mathbb{G}l_1 \times \mathbb{G}l_2\backslash\tilde{\mathbb{B}}_0 & \xrightarrow{\sim} & \mathbb{B}_0 \,. \end{array} \tag{2.60}$$

Proposition 2.31. *$\mathbb{B}_0$ is dense in $\mathbb{B}$. More precisely, $\mathbb{B}\backslash\mathbb{B}_0$ is a closed analytic subset of $\mathbb{B}$ of dimension ≤ 1.*

For the proof we need some basic facts of algebraic geometry. An algebraic variety X is called *quasiprojective*, if there exists an open algebraic embedding $X \hookrightarrow \bar{X}$, $\bar{X} \subset \mathbb{P}^N$ a projective variety. We assume that our embedding is an inclusion. Then $\bar{X}\backslash X$ is a closed subvariety of X, and $\bar{X}$ the ZARISKI closure of X in $\mathbb{P}^N$. There is a decomposition

$$\bar{X}\backslash X = \bigcup_{i=1}^{n} Y_i \ , \ Y_i \text{ closed irreducible subvariety of } X.$$

Now we can consider the space

$$\tilde{\mathbb{H}}_g = \{\Pi \in \mathrm{MAT}_{g\times 2g}(\mathbb{C}) \ ; \quad \Pi I^t\Pi = 0 \, , \ \Pi J^t\bar{\Pi} < 0\} \tag{2.61}$$

of RIEMANN matrices (of size $g \times 2g$). The columns of Π generate a lattice $\wedge_\Pi \subset \mathbb{C}^g$ and $A = \mathbb{C}^g/\wedge$ is an abelian variety, see Theorem 2.6. We call the pair (A, Π) a *labeled abelian variety*. In this sense $\tilde{\mathbb{H}}_g$ is the space of labeled abelian varieties (of dimension g).

We call two labeled abelian varieties (A, Π) and (A', Π') equivalent, if there is a commutative diagram

$$\begin{array}{rcccc} & \mathbb{C}^g & \longrightarrow & \mathbb{C}^g/\wedge_\Pi & \cong A \\ G\cdot & \downarrow\wr & & \downarrow\wr & \downarrow\wr \\ & \mathbb{C}^g & \longrightarrow & \mathbb{C}^g/\wedge_{\Pi'} & \cong A' \end{array}$$

with vertical isomorphisms, $G \in \mathbb{G}l_g(\mathbb{C})$, such that $\Pi' = G\Pi$. The pairs $(A, \mathbb{G}l_g(\mathbb{C})\cdot \Pi)$, $A \cong \mathbb{C}^g/\wedge_\Pi$, are called *marked abelian varieties.* We recover the generalized SIEGEL upper half space $\mathbb{H}_g = \mathbb{G}l_g(\mathbb{C})\backslash\tilde{\mathbb{H}}_g$ as "moduli space of marked abelian varieties" of dimension g (space of period points in section 2).

Let $E : \wedge_\Pi \times \wedge_\Pi \longrightarrow \mathbb{Z}$ be a principal polarization, see section 2. It is determined by the values of pairs of base vectors of $\wedge_\Pi$, hence by the E-values of pairs of the columns of Π. Denote the columns by $\mathfrak{a}_1, \dots, \mathfrak{a}_g, \mathfrak{b}_1, \dots, \mathfrak{b}_g$. Assume that this is a FROBENIUS (or symplectic) basis of $\wedge_\Pi$, that means that

$$E(\mathfrak{a}_i, \mathfrak{a}_j) = E(\mathfrak{b}_i, \mathfrak{b}_j) = 0 \,, \; E(\mathfrak{a}_i, \mathfrak{b}_j) = \delta_{ij} \,, \quad 1, \dots, g \,.$$

FROBENIUS bases are uniquely determined by E up to the action of the symplectic modular group $\mathbb{S}p(2g, \mathbb{Z})$. So principally polarized abelian varieties (A, E) can be identified with bi-equivalence classes of labeled abelian varieties (A, Π) with respect to $\mathbb{G}l_g(\mathbb{C})$ and $\mathbb{S}p(2g, \mathbb{Z})$ or, in other words, with pairs $(A, \mathbb{G}l_g(\mathbb{C})\Pi\mathbb{S}p(2g, \mathbb{Z}))$.

Definition-Proposition 2.32. *The moduli space of (principally) polarized varieties of dimension g is the quasiprojective variety*

$$\mathcal{A}_g = \mathbb{G}l_g(\mathbb{C})\backslash\tilde{\mathbb{H}}_g/\mathbb{S}p(2g, \mathbb{Z}) = \mathbb{H}_g/\mathbb{S}p(2g, \mathbb{Z}) \,.$$

Example 2.33.

$$\mathcal{A}_1 = \mathbb{H}/\mathbb{S}p(2, \mathbb{Z}) = \mathbb{H}/\mathbb{S}l_2(\mathbb{Z}) = \mathbb{P}^1\backslash\{\infty\}$$

is the moduli space of elliptic curves.

Now we change back to curves of genus g. The set of isomorphy classes of smooth curves of genus g can be endowed with the structure of a quasiprojective variety. We denote it by $\mathcal{M}_g$. Any algebraic compactification of $\mathcal{M}_g$ is called *moduli space of curves* of genus g, and $\mathcal{M}_g$ itself we will call the *moduli space of smooth curves* of genus g. The next theorem goes back to TORELLI (see Theorem 2.11).

Theorem 2.34. *For $g \geq 1$ there is an algebraic embedding*

$$jac : \mathcal{M}_g \longrightarrow \mathcal{A}_g \,, \quad cl(C) \longmapsto cl(Jac\, C)$$

corresponding to each isomorphy class of a (smooth) curve the isomorphy class of its canonically polarized Jacobian variety $Jac\, C = (J(C), E_C)$.

Let $\hat{\mathcal{A}}_g$ be the (projective) BAILY-BOREL compactification of $\mathcal{A}_g$ and $\hat{\mathcal{M}}_g$ the closure of the image $jac(\mathcal{M}_g)$ in $\hat{\mathcal{A}}_g$. Then $\mathcal{M}_g \to \hat{\mathcal{M}}_g$ is an open embedding.

Sometimes it is useful to work with a compactification $\bar{\mathcal{A}}_g$ of $\mathcal{A}_g$ such that the compactification cycle $\bar{\mathcal{A}}_g\backslash\mathcal{A}_g$ is a divisor with normal crossings. Such compactifications exist by the Theorem of HIRONAKA on resolutions of singularities. In this case the morphism jac can be extended to a morphism $\overline{jac} : \bar{\mathcal{M}}_g \to \hat{\mathcal{A}}_g$ by a theorem of BOREL [10]. We call jac the TORELLI *embedding*, $\overline{jac}$ a TORELLI morphism and the rational map of any algebraic compactification of $\mathcal{M}_g$ into any algebraic compactification of $\mathcal{A}_g$, induced by jac, a TORELLI *map*.

Now remember the moduli space $\mathbf{M}^0$ of smooth PICARD curves classified already in Proposition 2.3. We have a composition of embeddings

$$(\mathbb{P}^2\backslash \triangle\!\!\!\star\,)/S_4 = \mathbf{M}^0 \longrightarrow \mathcal{M}_3 \underset{jac}{\longrightarrow} \mathcal{A}_3\ , \tag{2.62}$$

where the first one is closed. Now we can establish a big commutative diagram (2.63) containing a "uniformization" (lift to symmetric domains) of the (restricted) TORELLI embedding (2.62) in an explicit manner.

$$\begin{array}{ccccc}
\tilde{B}_0 & \hookrightarrow & \tilde{\mathbb{B}} & \xrightarrow{\ *\ } & \tilde{\mathbb{H}}_3 \\
\downarrow & & \mathbb{G}l_1\times\mathbb{G}l_2\backslash\downarrow & & \mathbb{G}l_3\backslash\downarrow \\
\mathbb{B}_0 & \hookrightarrow & \mathbb{B} & \xrightarrow{\ *\ } & \mathbb{H}_3 \\
\downarrow/\Gamma & & \Gamma\cong\downarrow\mathbb{S}p'(6,\mathbb{Z}) & & \downarrow\mathbb{S}p'(6,\mathbb{Z}) \\
(\mathbb{P}^2\backslash\triangle\!\!\!\star\,)\ /S_4=\mathbf{M}^0 & \hookrightarrow & \mathcal{A}_3' & \hookrightarrow & \mathcal{A}_3 \\
\updownarrow & & \| & & \\
\mathbb{B}_0/\Gamma & \hookrightarrow & \mathbb{B}/\Gamma & \hookrightarrow & \widehat{\mathbb{B}/\Gamma}
\end{array} \tag{2.63}$$

The upper part comes from diagram (2.60). The subgroup $\mathbb{S}p'(6,\mathbb{Z})$ has been defined in 2.25. The isomorphy with the PICARD modular group $\Gamma = \mathbb{U}((2,1),\mathcal{O})$ of EISENSTEIN numbers comes from Lemma 2.27 with explicit correspondence (2.50). So the middle part of (2.63) comes from arithmetic quotients. It defines the closed subvariety $\mathcal{A}_3'$ of the moduli space of principally polarized abelian varieties $\mathcal{A}_3$, and $\mathbb{B}_0/\Gamma$ has to be the (open) moduli (sub-)space (of $\mathcal{A}_3'$) of smooth PICARD curves; compare with (2.54).

Proof of Proposition 2.31: Since the algebraic embedding $\mathbf{M}^0 \subset \mathcal{A}_3'$ is open, the complement $(\mathbb{B}/\Gamma)\backslash(\mathbb{B}_0/\Gamma)$ is algebraically closed in $\mathbb{B}/\Gamma$, hence an analytic subset (of dimension ≤ 1). The quotient map $\mathbb{B}\to\mathbb{B}/\Gamma$ is locally finite. Now the statement of Proposition 2.31 follows immediately.

The precise determination of $\mathbb{B}_0 \subset \mathbb{B}$ needs finer methods. We refer to the next chapter for this purpose. Assume that this is done. Then we have solved the first three problems of the effective SCHOTTKY-TORELLI problem 2.12. Namely, assume that the smooth curve C of genus 3 has a typical period matrix Π described in Lemma 2.21. Then it belongs to $\tilde{\mathbb{B}}_0$ introduced at the beginning of this section. The diagram (2.63) teaches us that the moduli point of $Jac(C)$ lies in the open part $\mathbb{B}_0/\Gamma$ of $\mathcal{A}_3'$. Therefore there is a smooth PICARD curve C' with the same moduli point $cl(Jac(C'))$. By TORELLI's Theorem 2.34 there is an isomorphism $C\cong C'$. This means that C itself is a PICARD curve. Especially, the smooth PICARD curves are the only smooth curves of genus 3 supporting a typical H_1-basis, which can be generally defined by producing a typical period matrix described in Lemma 2.21. Observe in this context that the cycle configurations (2.37) are "typical" for PICARD curves. It explains our nomenclature.

Furthermore, we are able to recover a typical period matrix from its three entries A_1, A_2, A_3 placed in Π as in (2.28) up to an elementary ambiguity. For this purpose we set $\mathfrak{a} = (A_1, A_2, A_3)$. Then we choose a basis $\mathfrak{b}, \mathfrak{e}$ of $\mathfrak{a}^\perp$ and find a typical period matrix $\begin{pmatrix} *\mathfrak{a} \\ \hline *\mathfrak{b} \\ \hline *\mathfrak{e} \end{pmatrix}$, if $\langle \mathfrak{a}, \mathfrak{a} \rangle < 0$ by Lemma 2.22, or, more precisely, if $\mathfrak{a}$ belongs to $\mathbb{B}_0$. The ambiguity comes from the action of $\mathbb{G}l_2(\mathbb{C})$ on $\mathfrak{a}^\perp$. Observe that the action of $\mathbb{B}((2,1),\mathcal{O})$ described in (2.50) does not play an additional role here because $\mathfrak{a}$ has to be stable.

The condition $\langle \mathfrak{a}, \mathfrak{a} \rangle < 0$ we looked for in 2.12.3 is called the *ball criterion.*

Remark 2.35. The image of $\mathbb{B}_0$ in $\mathbb{H}_3$, see diagram (2.63), can be explicitely described. This is a simple exercise of linear algebra. One has only to use the relations between the vectors in the typical period matrices $\Pi = (\Pi_1|\Pi_2)$ described in Lemma 2.22 in order to find the "typical period points" Ω by calculating $\Pi_1^{-1} \cdot \Pi = (E_3|\Omega)$. PICARD carried out this calculation in [60]. With the notations of Lemma 2.21 and (2.45) the result is

$$\Omega = \begin{pmatrix} (\bar{\rho}u^2 + 2\rho v)/(1-\bar{\rho}), & -\bar{\rho}u, & (u^2 - \rho v)/1 - \bar{\rho}) \\ -\bar{\rho}u, & \bar{\rho}, & -u \\ (u^2 - \rho v)/(1-\bar{\rho}), & -u, & (\rho u^2 + 2\rho v)/(1-\bar{\rho}) \end{pmatrix} \tag{2.64}$$

with $u = A_2/A_1$, $v = \bar{\rho}A_3/A_1$ (A_1 cannot be equal to 0). The ball criterion $\langle \mathfrak{a}, \mathfrak{a} \rangle < 0$ transfers to

$$2Re(v) + |u|^2 < 0\,. \tag{2.65}$$

3 Uniformizations and Differential Equations of Euler-Picard Type

3.1 Ball Uniformization of Algebraic Surfaces

Let X be a normal complex algebraic surface. We assume for a moment that X is compact. Then it supports only a finite number of singularities. Furthermore we assume that all these singularities are of quotient or ball cusp type.

Definition 3.1. A surface germ (U, P), U an open analytic surface (neighbourhood of P), $P \in U$, is called a *quotient singularity*, if (U, P) is the finite quotient $(V, O)/G$ of a smooth germ (V, O).

This means that there is a finite group G acting on a smooth analytic surface V, G fixes the point $O \in V$, $U \cong V/G$ and P corresponds to the quotient image of 0 under this isomorphism. The quotient singularities have been completely classified by means of resolution graphs in [14]. Looking at the tangent space of V at O one gets all quotient singularities as surface germs $(\mathbb{C}^2/G, \bar{O})$, G a finite subgroup of $\mathbb{G}l_2(\mathbb{C})$, $\bar{O}$ the image of $O = (0,0) \in \mathbb{C}^2$. If G is abelian, then (U, P) is called an *abelian (quotient) singularity*.

A more algebraic approach can be given by blowing up the 0-point of $\mathbb{C}^2$ in the above situation. More generally, let $D = \mathbb{P}^1$ and L/D be an (affine) line bundle over D, that means the fibres of the (affine) projection $L \to D$ are isomorphic to the affine line $\mathbb{A}^1 = \mathbb{A}^1(\mathbb{C})$. The (image of the) zero section is also denoted by D. We consider only negative line bundles defined by the condition $(D^2) < 0$, where

$$(D^2) = (D^2)_L = \textit{selfintersection number} \text{ of } D \text{ in } L.$$

Let G be a finite group acting (algebraically) on L/D. Then we can contract D/G on L/G to a point P and $(\widehat{L/G}, P)$ is a quotient singularity. In this way one gets all quotient singularities.

The same can be done with negative linebundles L/D for arbitrary smooth compact curves D.

Definition 3.2. If in the above construction $D = E$ is an elliptic curve, then we call $(\widehat{L/G}, P)$ *a ball cusp singularity.*

The ball cusp singularities (in dimension 2) have been first classified by the author, see e.g. [32]. It turns out that for each of them there is a local ball lattice $\Gamma \subset \mathbb{U}((2,1), \mathbb{C})$ fixing a boundary point $\kappa \in \partial\mathbb{B}$ of $\mathbb{B}$, a Γ-invariant open analytic subset V of $\mathbb{B}$ with boundary point (cusp) κ, such that

$$\widehat{V/\Gamma} = V \cup \{\kappa\}/\Gamma = (V/\Gamma) \cup \{\hat{\kappa}\}$$

has a canonical normal complex structure and the singularities $(\widehat{V/\Gamma}, \hat{\kappa})$ coincide with the cusp singularity we started with. Moreover, all ball cusp singularities can be presented in this way, see [32]. For a higher dimensional construction of (neat) ball cusp singularities we refer to [58].

We remark that there are ball cusp singularities, which are also quotient singularities. We want to distinguish them by introducing "weights" of the points and of some curves through them. More precisely, we consider tuples

$$(U; v_1 C_1, \ldots, v_k C_k; P) , \tag{3.1}$$

where (U, P) is a normal surface singularity (germ), $C_1, \ldots, C_k$ are smooth analytic curves (germs) on U going through P and $v_1, \ldots, v_k$ are positive natural numbers called *weights* of $C_1, \ldots, C_k$, respectively. For example, if $(U, P) \cong (\mathbb{C}^2/G, \bar{O})$, G a finite abelian subgroup of $\mathbb{G}l_2(\mathbb{C})$, then the quotient map $\mathbb{C}^2 \to U$ is a finite covering with (at most) two branch curves C_1, C_2 on U and well-defined ramification indices v_1, v_2. The corresponding quadruple $(U; v_1 C_1, v_2 C_1; P)$ (germ) is called an *abelian point.*

A *quasiresolution* of our tuple (3.1) is a weighted curve germ tuple $(\tilde{U}; vC, v_1 C_1, \ldots, v_k C_k)$ of our tuple, $v \in \mathbb{N}_+$, together with an analytic morphism $\sigma : \tilde{U} \to U$ such that the preimage of P is the compact smooth curve C, σ is isomorphic outside of C of P, respectively, $C_1, \ldots, C_k$ are the proper transforms of the curve germs on U through P with the same notations, and $(U; vC, v_i C_i; P_i)$, P_i the intersection point of C and C_i on U, is an abelian point for $i = 1, \ldots, k$. If a quasiresolution exists, then we say that the curve germs C_i *cross each other quasinormally* at P.

Definition 3.3. If the germ (3.1) has quasinormal crossings and is not an abelian point, then we call

$$v\mathbf{P} = (U; v_1 C_1, \ldots, v_k C_k; vP) , \quad v \in \mathbb{N}_+ \cup \{\infty\} , \tag{3.2}$$

an *orbital point* with *central weight* v. An abelian point is a *centrally weightless* orbital point.

Let G be a finite group acting on a line bundle L/D as above and C the image of D on L/G along $L \to L/G$. We correspond to C its ramification index v. The other branch curves $C_1, \ldots, C_k$ with branch indices $v_1, \ldots, v_k$, respectively, come from fibres of L/D. We can contract C to a point P and obtain in this way an orbital point (3.2) with finite central weight v.

Definition 3.4. A *quotient point* is an abelian point or an orbital point coming out by dividing a line bundle $L/\mathbb{P}^1$ by a finite group as described above. If we take an elliptic base curve $D = E$ and define the central weight to be $v = \infty$, then the arising orbital point is called a *(ball) cusp point.*

A precise classification list with resolution graphs and weights can be found in [39]. If Γ is an arithmetic subgroup of $\mathbb{U}((2,1),\mathbb{C})$, then the BAILY-BOREL compactification $\widehat{\mathbb{B}/\Gamma}$ has only quotient and cusp singularities. By a theorem of BOREL [9] there is a freely acting cofinite normal subgroup Γ' of Γ. The factor group Γ/Γ' acts on the normal compact algebraic surface $\widehat{\mathbb{B}/\Gamma'}$ and we can use the ramification indices of the corresponding finite GALOIS covering $\widehat{\mathbb{B}/\Gamma'} \to \widehat{\mathbb{B}/\Gamma}$ to endow curves through the singularities of the branch locus with the corresponding weights. The central weights can be obtained in the same manner after blowing up the preimages of $\widehat{\mathbb{B}/\Gamma'}$ of non-abelian points on $\widehat{\mathbb{B}/\Gamma}$. The weights, hence all quotient points lying on the "finite part" $\mathbb{B}/\Gamma$, do not depend on the choice of Γ'. But it is not possible in a canonical way to define a finite central weight at the compactification points. Therefore we are well-motivated and it is quite sufficient for our purposes to set these weights equal to ∞.

Definition 3.5. An *orbital curve* is a triple

$$\mathbf{C} = (U, vC, \sum_i v_i \, \mathbf{P}_i)\,, \quad v, v_i \in \mathbb{N}_+ \tag{3.3}$$

where (U, C) is an analytic surface germ along a compact curve C, the $v_i \, \mathbf{P}_i$'s are orbital points on (U, vC), this means that the weighted curve germ (U, vC) appears as first of the weighted curve germs through P_i in the definition of orbital points, see (3.2), if C is smooth at P_i. If not, then each of the (smooth) branches of C, through P_i have to appear with the same weight v.

Furthermore it is assumed that all surface singularities of U on C and all curve singularities of C belong to the supporting point set $\{P_1, \ldots, P_r\}$ of the *orbital curve cycle* $\Sigma v_i \mathbf{P}_i$.

For example, each quasiresolution of a finite orbital point yields an orbital curve with only abelian points in the corresponding orbital cycle. The same is true for the quasiresolution of a cusp point, but with central curve weight $v = \infty$. We allow this extension of definition and call such quasiresolutions (with $v = \infty$) *orbital cusp curves.*

An *orbital cycle* on a surface X is a formal sum of orbital curves on X such that each intersection point P of the supporting irreducible components is the support of an orbital point on each of the orbital components going through P. Formally an orbital cycle $\mathbf{D}$ on X can be written as

$$\mathbf{D} = \Sigma v_i C_i + \Sigma w_j \mathbf{P}_j\,, \quad v_i, w_j \in \mathbb{N}_+ \cup \{\infty\}\,. \tag{3.4}$$

Weightless abelian points have been omitted. We admit only those which can be discovered automatically at intersection points of (precisely) two irreducible curve components of D or at abelian (cyclic) surface singularities on (precisely) one of these components. Moreover, we claim that all quotient and cusp singularities belong to the point part of (3.4).

Definition 3.6. A pair $\mathbf{X} = (X, \mathbf{D}), \mathbf{D}$ an orbital cycle on X, is called an *orbital surface.*

Example 3.7. Let $X = \mathbb{P}^2$, Q_1, Q_2, Q_3, Q_4 four points on $\mathbb{P}^2$ in general position, $L_{ij} = L_{ji}$ the line through Q_i, Q_j. With the notations of (3.4) we define an orbital cycle by $\mathbf{D} = \sum_{1 \le i < j \le 4} 3L_{ij} + \sum_{k=1}^{4} \infty \cdot \mathbf{Q}_k$. The pair $(\mathbb{P}^2, \mathbf{D})$ is an orbital surface.

Remark 3.8. If X is a multi-blowing up of $\mathbb{P}^2$ and $\mathbf{D}$ is supported by lines, then HIRZEBRUCH called $\mathbf{D}$ a (linear) arrangement, see [5].

If we omit in D and on X all infinitely weighted curves and points, then we get, by definition, an *open orbital surface* $\mathbf{X}_f = (X_f, \mathbf{D}_f)$. The index symbol f comes from "finite part".

Main Example 3.9. Let Γ be an arithmetic subgroup of $\mathbb{U}((2,1), \mathbb{C})$. Then the quotient $X_f = \mathbb{B}/\Gamma$ is an (in general) open algebraic surface. The irreducible curves of the branch locus of the covering $\mathbb{B} \longrightarrow \mathbb{B}/\Gamma$ and the images of Γ-elliptic points on $\mathbb{B}$ are endowed with the corresponding ramification indices as weights. Denote the corresponding (open) orbital cycle by $\mathbf{D}_f$. Then $\mathbf{X}_f = (X_f, \mathbf{D}_f)$ is an open orbital surface. We add the infinitely weighted compactifying points of the BAILY-BOREL compactification $\widehat{\mathbb{B}/\Gamma}$ to the "closure" $\hat{D}_f$ of D_f. The pairs $(X_f, \mathbf{D}_f)$ or $(\hat{X}_f, \hat{\mathbf{D}}_f)$ are called *open* or *compactified orbital ball quotient surfaces*, respectively. Sometimes they are denoted by $\mathbb{B}/\mathbf{\Gamma}$ or $\widehat{\mathbb{B}/\mathbf{\Gamma}}$, respectively.

A fundamental question is, which orbital surfaces are ball quotients? A precise answer can be considered as a special but important part of the 22nd HILBERT PROBLEM entitled "Uniformization of analytic relations by means of automorphic functions".

In order to find a constructive answer we introduced in [39] successively some invariants called (*local* and *global*) *orbital heights* (respectively):

$$h_e, h_\tau : \{\text{orbital curves}\} \longrightarrow \mathbb{Q} \text{ (local heights)},$$

$$H_e, H_\tau : \{\text{orbital surfaces}\} \longrightarrow \mathbb{Q} \text{ (global heights)}.$$

They are called EULER *heights* (index e) or *signature heights* (index τ), respectively. We give an implicit definition:

(i) If every (irreducible) thing is smooth, compact and has trivial weights 1, then

$$\begin{aligned} h_e(\mathbf{C}) = h_e(C) &= -(\text{EULER number of } C), \\ h_\tau(\mathbf{C}) = h_\tau(C) &= -(\text{selfintersection index of } C \subset X), \\ H_e(\mathbf{X}) = H_e(X) &= \text{EULER number of } X, \\ H_\tau(\mathbf{X}) = H_\tau(X) &= \text{signature of } X. \end{aligned}$$

(ii) If $\mathbf{f} : \mathbf{X} \to \mathbf{X}', \mathbf{C} \to \mathbf{C}'$ are finite coverings, then the following degree formulas hold:

$$H(\mathbf{X}) = \deg(\mathbf{f})H(\mathbf{X}'), \quad h(\mathbf{C}) = \deg(\mathbf{f}|\mathbf{C}) \cdot \mathbf{h}(\mathbf{C}')\,,$$
$$H = H_e \text{ or } H_\tau, \qquad h = h_e \text{ or } h_\tau\,.$$

Roughly spoken, a *finite covering* $\mathbf{f}$ is a usual finite covering $f : X \to X'$ compatible with weights in the sense of GALOIS theory. The definition restricts to orbital curves.

Proofs of existence via explicit definitions and calculations can be found in [36], [37] and the related literature.

Theorem 3.10 ([39]). *If* $\mathbf{X} = (X, \mathbf{D}) = \widehat{\mathbb{B}/\mathbf{\Gamma}}$ *is an orbital ball quotient, then it holds that*

$$\begin{aligned} &\text{(i)} \qquad H_e(\mathbf{X}) = 3H_\tau(\mathbf{X}) > 0\,, \\ &\text{(ii)} \qquad h_e(\mathbf{C}) = 2h_\tau(\mathbf{C}) > 0 \end{aligned} \tag{3.5}$$

for all orbital curve components $\mathbf{C}$ *of* $\mathbf{D}$.

The discrete subgroup $\mathbb{P}\Gamma$ *of* $\mathbb{PU}((2,1),\mathbb{C})$ *is uniquely determined up to conjugation by the orbital surface* $\mathbf{X}$.

So we found rather effective necessary conditions for an orbital surface to be a ball quotient. R. KOBAYASHI [43] gave sufficient conditions in another language. Until now there is no proof of the equivalence of KOBAYASHI's and our conditions. The following considerations indicate that they cannot be far away from each other.

The relations (3.5) can be understood as a system $DIOPH(X, D)$ of diophantine equations and inequalities, if we write it down explicitly with all weights as variables. The coefficients depend only on geometric data of the supporting surface X and the supporting cycle D. It turns out that

Theorem 3.11 ([39]). *For any (admissible) pair* (X, D) *the system* $DIOPH(X, D)$ *has at most finitely many solutions. These solutions can be calculated in an effective manner.*

"Admissible" means that X has only quotient and cusp singularities and the singularities of the cycle D are simple enough to be supports of quotient or cusp points classified in [39].

So a ball uniformization of a surface X with given branch locus D can only happen, if we find a solution of $DIOPH(X, D)$, and there are, up to isomorphy, at most finitely many possibilities.

Example 3.12. If we fix infinite weights at the four triple points of  $\subset \mathbb{P}^2$ defined in (2.7), then $DIOPH(\mathbb{P}^2,$ $)$ has exactly one solution. The corresponding orbital surface $(\mathbb{P}^2, \mathbf{D})$ coincides with that of Example 3.7.

The corresponding calculations for this simple example can be already found in [5]. Now one could check KOBAYASHI's conditions to see that the ball uniformization of $(\mathbb{P}^2, \mathbf{D})$ really exists. In [33] we proved more. There we presented the corresponding uniformizing ball lattice explicitly:

Proposition 3.13. *Let* $\Gamma(\sqrt{-3}) \subset \mathbb{U}((2,1),\mathcal{O})$, $\mathcal{O} = \mathcal{O}_K$, $K = \mathbb{Q}(\sqrt{-3})$, *be the congruence subgroup corresponding to the ideal* $(1-\rho)$ *of* $\mathcal{O}$. *Then the orbital surfaces* $\widehat{\mathbb{B}/\mathbf{\Gamma}}(\sqrt{-3})$ *and* $(\mathbb{P}^2, \mathbf{D})$ *coincide.*

The congruence subgroup $\Gamma(\sqrt{-3})$ is defined by the exact sequence

$$\begin{array}{c} 1 \longrightarrow \Gamma(\sqrt{-3}) \longrightarrow \Gamma = \mathbb{U}((2,1),\mathcal{O}) \longrightarrow \mathbb{U}((2,1),\ \mathcal{O}/\sqrt{-3}\mathcal{O})) \longrightarrow 1 \\ \hspace{9em} \| \wr \\ \hspace{9em} S_4 \times Z_2 \end{array} \tag{3.6}$$

For the proof of Proposition 3.13 it was necessary to use the following two theorems.

Theorem 3.14 ([39]). *For the* c_2*-volume* $c_2(\Gamma)$ *of a fundamental domain of a ball lattice* Γ *with respect to the* BERGMANN *metric on* $\mathbb{B}$ *it holds that*

$$c_2(\Gamma) = H_e(\widehat{\mathbb{B}/\mathbf{\Gamma}}), \quad c_2(\Gamma)/3 = H_\tau(\widehat{\mathbb{B}/\mathbf{\Gamma}}). \tag{3.7}$$

In [38] we presented an effective formula for $c_2(\Gamma_M)$, $\Gamma_M = \mathbb{U}((2,1),\mathcal{O}_M)$ the full PICARD modular group of the imaginary quadratic number field M, in terms of a special value of L-series using arithmetic-geometric methods in the proof. It turns out that

Theorem 3.15 ([38]).

$$c_2(\Gamma_M) = \delta \cdot L\left(3, \left(\frac{D_{M/\mathbb{Q}}}{\bullet}\right)\right) \cdot |D_{M/\mathbb{Q}}|^{5/2}/32\pi^3,$$

where $\delta = 3$, *if* $M = K = \mathbb{Q}(\sqrt{-3})$, *or* $\delta = 1$ *otherwise;* $D_{M/\mathbb{Q}}$ *is the discriminant of* $M/\mathbb{Q}$ *and* L *denotes the* L-series *defined by*

$$L\left(s, \left(\frac{D}{\bullet}\right)\right) = \sum_{n=1}^{\infty} \left(\frac{D}{n}\right)/n^s, \quad D = D_{M/\mathbb{Q}}.$$

With these formulas one obtains successively

$$H_e(\widehat{\mathbb{B}/\mathbf{\Gamma}}(\sqrt{-3})) = 1/3, \ H_\tau(\widehat{\mathbb{B}/\mathbf{\Gamma}}(\sqrt{-3})) = 1/9,$$

the CHERN numbers

$$c_2(\widehat{\mathbb{B}/\Gamma}(\sqrt{-3}) = 3, \quad c_1^2(\widehat{\mathbb{B}/\Gamma}(\sqrt{-3}) = 9$$

after classification of elliptic points and cusps, and finally $\widehat{\mathbb{B}/\Gamma}(\sqrt{-3}) = \mathbb{P}^2$ by the theory of surface classification.

The classification of $\widehat{\mathbb{B}/\Gamma}(\sqrt{-3})$ can be found in all details in [33]. There has been also determined precisely the ramification locus of the "uniformization" $\mathbb{B} \longrightarrow \mathbb{B}/\Gamma(\sqrt{-3}) = \mathbb{P}^2 \setminus \{Q_1, Q_2, Q_3, Q_4\}$:

Corollary 3.16. *The preimage of* $\bigtriangleup\!\!\!\!\star \subset \mathbb{P}^2$ *on* $\mathbb{B}$ *is the infinite union*

$$\Diamond\!\!\!\!+ = \Gamma\mathbb{D} = \{\gamma\mathbb{D}; \gamma \in \Gamma\}, \quad \mathbb{D} : A_2 = 0, \ \ \Gamma = \Gamma_K , \tag{3.8}$$

of subdiscs of $\mathbb{B}$, *with the notations of* (2.48).

The symbol $\Diamond\!\!\!\!+$ comes from six "generating discs" with respect to $\Gamma(\sqrt{-3})$, namely, one can prove that $\Gamma\mathbb{D}$ is the $\Gamma(\sqrt{-3})$-orbit of six discs on the standard ball $\mathbb{B} \subset \mathbb{C}^2$ lying on the six lines through pairs of the four points $(0, 1)$, $(1, 0)$, $(0, -1)$, $(-1, 0) \in \partial\mathbb{B}^2$, see [33].

Now look back to diagram (2.59). The objects and morphisms of the right triangle are determined: $\Gamma = \Gamma_K$, $\Gamma' = \Gamma(\sqrt{-3})$ and the point set {points} consists of the four triple points of $\bigtriangleup\!\!\!\!\star$.

We are also able to fill a gap at the end of 2.7. From Corollary 3.16 it follows that

$$\mathbb{B}_0 = \mathbb{B} \backslash \Diamond\!\!\!\!+ \tag{3.9}$$

So, up to linear algebra (see diagram (2.63)), we determined in this way precisely the space $\mathbb{B}_0$ of typical period matrices of smooth PICARD curves. Altogether we accomplished the fine solution of the relative SCHOTTKY problem for PICARD curves 2.12, 1.–31.

The next section prepares the understanding of (diagram 2.59) in terms of analytic functions.

3.2 Special Fuchsian Systems and GAUSS-MANIN Connection

In [91] M. YOSHIDA succeeded to solve a higher-dimensional version of the RIEMANN-HILBERT problem. The background is HILBERT's 21st Problem "Proof of the existence of linear differential equations with prescribed monodromy groups" set up for functions of one variable, "... to show that in any case there exists a linear differential equation of the Fuchsian class with given singular locus and prescribed monodromy group". The final solution of this HILBERT Problem has been given by H. ROEHRL [64]. In the following we shall restrict ourselves to the second dimension.

Theorem 3.17 (M. YOSHIDA [91]). *Let* $\mathbf{X} = (X, \mathbf{D})$ *be an orbital surface uniformizable by the ball with quotient map* $p : \mathbb{B} \to X_f$. *Then the inverse* p^{-1} *of* p *is a (multivalued) developing map of a Fuchsian system of linear partial differential equations.*

It means that there is locally a fundamental system of solutions I_0, I_1, I_2 extending analytically to $X\backslash D$, D the support of $\mathbf{D}$, such that the multivalued map

$$(I_0 : I_1 : I_2) : X\backslash D \longrightarrow \mathbb{B} \subset \mathbb{P}^2 ,$$

$P \longmapsto (I_0(P) : I_1(P) : I_2(P))$, coincides with p^{-1} on $X\backslash D$. The Fuchsian system is called the *uniformizing equation* of the orbital surface and the uniformizing ball lattice $\Gamma \subset \mathbb{PU}((2,1), \mathbb{C})$ is the *monodromy group* of the system. Via solutions one gets a unitary representation of the fundamental group,

$$\pi_1(X\backslash D) \longrightarrow \mathbb{P}\Gamma \subset \mathbb{PG}l_3(\mathbb{C}) .$$

YOSHIDA [91] found an effective method in order to determine a corresponding Fuchsian system. Together with Proposition 3.13 one gets

Theorem 3.18. *The following Fuchsian system* (3.10) *of partial differential equations*

$$\begin{aligned} D_{ij}F(u,v) &= 0 \text{ on } \mathbb{C}^2\backslash\boxslash = \mathbb{P}^2\backslash\bigstar\!\!\!\!\bigtriangleup , \qquad (3.10)\\ (i,j) &= (1,1),(1,2),(2,2) ,\\ D_{11} &= \frac{\partial^2}{\partial u} + [9(u-1)u(v-u)]^{-1} \cdot \\ &\cdot \left\{ 3(-5u^2+4uv+3u-2v)\frac{\partial}{\partial u} + 3(v-1)v\frac{\partial}{\partial u} + (u-v) \right\} ,\\ D_{12} &= \frac{\partial^2}{\partial u \partial v} + [3(u-v)]^{-1} \left\{ \frac{\partial}{\partial u} - \frac{\partial}{\partial v} \right\} ,\\ D_{22} &= \frac{\partial^2}{\partial v^2} + [9(v-1)v(u-v)]^{-1} \cdot \\ &\cdot \left\{ 3(-5v^2+4uv+3v-2u)\frac{\partial}{\partial u} + 3(u-1)u\frac{\partial}{\partial u} + (v-u) \right\} , \end{aligned}$$

is an uniformizing equation of the orbital surface $(\mathbb{P}^2, \underline{\bigstar\!\!\!\!\bigtriangleup})$. *Its monodromy group is the* PICARD *modular group* $\Gamma(\sqrt{-3})$.

YOSHIDA's general approach lifting the GAUSS-SCHWARZ theory of Fuchsian equations to higher dimensions has a classical origin in the work of PICARD and APPELL. Especially for linear arrangements of low-dimensional projective spaces a more immediate result known as *PTDM*-Theorem (due to PICARD, TERADA, DELIGNE, MOSTOW) would be sufficient for our purposes. We refer to [91], [5] and further literature given there. But we prefer to change over from the analytic viewpoint to an algebraic-geometric approach in order to find "algebraic solutions" of special Fuchsian equations represented by integrals on algebraic curves along cycles depending on parameters u, v. The general framework of the corresponding algebraic theory is known as GAUSS-MANIN *connection* of algebraic families of algebraic manifolds. We give a short introduction and outline for special algebraic curve families. For more details and proofs we refer to [33].

Our special purpose for the use of GAUSS-MANIN connections in [33] was to understand integrals of the form $\int q(x)dx/\sqrt[n]{p_m(x)}$ from an algebraic-geometric view point; $q(X)$ and $p_m(X)$ are polynomials in one variable X, p_m of degree m. Especially a variation of $p_m(X)$ yields analytic integral functions $\int\limits_\gamma q(x)dx/y^\ell$, taken on the (compactified, desingularized) RIEMANN surface $\tilde{C} : Y^n = p_m(X)$ along cycles γ, depending on the zeros of p_m. We assume that $q(x)dx/y^\ell$ is a differential form of second kind, that means that all residues vanish. Then the integral depends only on its cycle class in $H_1(\tilde{C}, \mathbb{Z})$ because the integral along small cycles around any point vanish in this case. Especially holomorphic differentials and *exact differentials* df, f a meromorphic function on $\tilde{C}$, are of second kind.

We look for Fuchsian systems of partial differential equations satisfied by our *(hypergeometric)* integral functions. One hopes to find effectively a fundamental system "generating" all of them in an algebraic manner such that there are no other solutions than the hypergeometric functions of fixed equation and differential type we started with. In order to be effective we will concentrate our attention to the case $q(X) = X^k$.

The first step of algebraization is done by POINCARÉ-DE RHAM duality: The cohomology group $H^1(\tilde{C}, \mathbb{C})$ is isomorphic to the hypercohomology group

$$\mathbb{H}^1(\Omega^\bullet_{\tilde{C}}) = \{\text{diff. forms of 2nd kind}\}/\{\text{exact diff. forms}\}.$$

We refer to [26]. It doesn't matter to work with meromorphic or with algebraic differential forms. The duality is a pairing described in the commutative diagram (3.11).

$$\begin{array}{ccc} H_1(\tilde{C}, \mathbb{C}) \times H^1(\tilde{C}, \mathbb{C}) & \longrightarrow & \mathbb{C} \\ \downarrow\| \qquad\qquad \downarrow\wr & & \\ H_1(\tilde{C}, \mathbb{C}) \times \mathbb{H}^1(\Omega^\bullet_{\tilde{C}}) & & \end{array} \tag{3.11}$$

extending

$$(\gamma, \omega) \longmapsto \int\limits_\gamma \omega\,.$$

Locally one can write a differential form of 2nd kind as sum of a holomorphic (first kind) form and an exact form. Let $P \in \tilde{C}$, ω a differential form of 2nd kind and U_1, U_2 (open) neighbourhoods of P such that $\omega|U_i = \omega_i + df$, $i = 1, 2$, ω_i of first kind on U_i, respectively. Setting $f_{12} = f_1 - f_2$ on $U_{12} = U_1 \cap U_2$, then it is clear that $\omega_1 - \omega_2 + df_{12} = 0$ on U_{12}. In this way each differential form of 2nd kind can be represented by a *hypercochain* $\{(\omega_i),(f_{ij})\}$ on an open covering $\{U_i\}$ of $\tilde{C}$. Hypercoboundaries are defined to come from exact differential forms such that the above pair is a *hypercocycle*, if $\omega_i - \omega_j + df_{ij} = 0$ for $i \neq j$.

The construction extends to higher dimensional (smooth) manifolds M starting with the DE RHAM complex

$$\Omega_M^\bullet : \mathcal{O}_M \xrightarrow{d} \Omega_M^1 \xrightarrow{d} \Omega_M^2 \xrightarrow{d} \dots .$$

By means of open coverings $\mathcal{U} = \{U_i\}$ and the ČECH co-differential maps δ defining coboundaries and cycles one gets ČECH-DE RHAM *bicomplexes*

$$\begin{array}{ccccccc}
C^0(\mathcal{U},\mathcal{O}_M) & \xrightarrow{d} & C^0(\mathcal{U},\Omega_M^1) & \xrightarrow{d} & C^0(\mathcal{U},\Omega_M^2) & \xrightarrow{d} & \dots \\
\delta\downarrow & & \delta\downarrow & & \downarrow & & \\
C^1(\mathcal{U},\mathcal{O}_M) & \xrightarrow{d} & C^1(\mathcal{U},\Omega_M^1) & \xrightarrow{d} & C^1(\mathcal{U},\Omega_M^2) & \xrightarrow{d} & \dots \\
\delta\downarrow & & \delta\downarrow & & \downarrow & & \\
C^2(\mathcal{U},\mathcal{O}_M) & \xrightarrow{d} & C^2(\mathcal{U},\Omega_M^1) & \xrightarrow{d} & C^1(\mathcal{U},\Omega_M^2) & \xrightarrow{d} & \dots \\
\delta\downarrow & & \delta\downarrow & & \delta\downarrow & &
\end{array}$$

Taking skew-diagonal direct sums in the bicomplex scheme above and a suitable new differential map combined by d and δ one obtains a *total* ČECH-DE RHAM *complex* with cohomology groups $H^i(C_{tot}^\bullet(\mathcal{U},\Omega_M^\bullet))$, see [42] for details. Taking limits over all coverings $\mathcal{U}$ the DE RHAM *hypercohomology groups* of M

$$\mathbb{H}_{DR}^i(M) = \mathbb{H}^i(\Omega_M^\bullet) = H^i\left(C_{tot}^\bullet(\Omega_M^\bullet)\right)$$

are well-defined. The construction works locally for open neighbourhoods of each point $P \in M$. The elements of $\mathbb{H}_{DR}(U)$ can be considered as sections of the *hypercohomology sheaf*

$$\underline{\mathcal{H}}_{DR}^i(M) = \underline{\mathcal{H}}^i(\Omega_M^\bullet) = \mathcal{H}^i(C_{tot}^\bullet) .$$

Now we change over to algebraic families $\mathcal{X}/S$ of algebraic varieties. We assume that everything is smooth, not only $\mathcal{X}$, S but also the morphism $\mathcal{X} \to S$ defining the family. Regarding differential forms on S as "trivial", one defines sheaves of relative differential forms $\Omega_{\mathcal{X}/S}^i \in \operatorname{mod} \mathcal{O}_S$ over S. There is a *relative* DE RHAM *complex* of $\mathcal{O}_S$-module sheaves

$$\Omega_{\mathcal{X}/S}^\bullet : \mathcal{O}_{\mathcal{X}} \longrightarrow \Omega_{\mathcal{X}/S}^1 \xrightarrow{d_{\mathcal{X}/S}} \Omega_{\mathcal{X}/S}^2 \longrightarrow \dots .$$

The construction of hypercohomology groups and sheaves via ČECH-DE RHAM bicomplexes works also in the relative case. We denote them by $\mathbb{H}_{DR}^i(\mathcal{X}/S)$ or $\underline{\mathcal{H}}_{DR}^i(\mathcal{X}/S)$, respectively.

Of our main interest are the families $\tilde{C}/T$ of curves

$$\begin{aligned} \tilde{C}_t : Y^n &= (X-1)X(X-t_1)\cdot\ldots\cdot(X-t_r)\,, \\ T &= \mathbb{C}^r \setminus \{t=(t_1,\ldots,t_r)\,;\quad t_i \neq t_j, 0, 1 \text{ for } i\neq j\}\,. \end{aligned} \tag{3.12}$$

Moving with t the differential forms

$$\omega = x^k dx/y^\ell \tag{3.13}$$

represent now a relative hypercohomology class

$$\bar{\omega} \in \mathbb{H}^1_{DR}(\tilde{C}/T)\,. \tag{3.14}$$

Thanks to nice base change properties proved by DELIGNE (see [18]) one recovers the differential form ω_t on $\tilde{C}_t$ given also by the expression in (3.13) from the global one. Namely, after localization and reduction, one gets

$$\mathbb{H}^1_{DR}(\tilde{C}/T)(t) = \mathbb{H}^1_{DR}(\tilde{C}_t)\,.$$

For a small open analytic subset U of T and cycle families $\underline{\gamma}/U = \{\gamma_t\}$ over U the analytic function

$$I_{\underline{\gamma}} = \int_{\underline{\gamma}/U} \bar{\omega} \text{ with values } I_{\underline{\gamma}}(t) = \int_{\underline{\gamma}_t} \omega_t\,, \quad t\in U\,, \tag{3.15}$$

is well-defined. In this way we introduced integrals over hypercohomology classes.

Remember that we look for differential operators D killing the functions (3.15). We would like to write the corresponding vanishing condition as

$$D\int_{\underline{\gamma}} \bar{\omega} = \int_{\underline{\gamma}} D\bar{\omega} = 0\,. \tag{3.16}$$

For this purpose one has to explain "differentiation of hypercohomology classes". This has been first done by MANIN [51] and later generalized and globalized essentially by KATZ, GROTHENDIECK and DELIGNE (see e.g. [42], [18]). It turns out that the hypercohomology sheaves $\underline{\mathcal{H}}^i_{DR}(\mathcal{X}/S)$ are not only $\mathcal{O}_S$-modules but also $\mathcal{D}_S$-modules, where $\mathcal{D}_S$ is the non-commutative ring sheaf of differential operators on S. For example, $\mathcal{D}_T$ is locally generated by regular functions and the coordinate derivations $\partial/\partial t_i$, $i=1,\ldots,r$.

For explaining such derivations one uses "relative étal coverings" $\{\mathcal{U}_i/S\}$ of $\mathcal{X}/S$. These are open coverings $\mathcal{U}_i$ of $\mathcal{X}$ having a decomposition $\mathcal{U}_i \to \mathbb{A}^k_S \to S$, where the first morphism is étal over its image, $\mathbb{A}^k_S = \mathbb{A}^k \times S$.

For example, in our curve family (3.12) we work mainly with

$$\mathcal{U}_0 = \mathcal{Y} = \left\{ (x, y; t) \in \mathbb{C}^2 \times T \,;\; y^n = \prod_{i=-1}^{r} (x - t_i)\,, \quad x \neq t_i \text{ for all } i \right\}$$

setting $t_{-1} = 1$, $t_0 = 0$. Then we obtain a relative étal map $\mathcal{Y} \to \mathbb{A}^1_T \to T$, $(x, y; t) \mapsto (x; t) \mapsto t$, involving an n-cyclic unramified GALOIS covering $\mathcal{Y} \to \mathbb{A}'_T$, $\mathbb{A}'_T \subset \mathbb{A}^1_T$.

Since $\mathcal{Y}$ and T are affine it holds that $\mathbb{H}^1_{DR}(\mathcal{Y}/T) = H^1_{DR}(\mathcal{Y}/T)$. On the coordinate ring

$$\mathbb{C}[T] = \mathbb{C}\left[t_1, \ldots, t_r, \prod_{i<j} (t_i - t_j)^{-1} \right]$$

of T the derivations $D_i = \partial/\partial t_i$, $i = 1, \ldots, r$ are well-defined. We extend them to $\mathbb{C}[\mathcal{Y}]$ and $\Omega^1_{\mathcal{Y}/T}$ by $D_i : x \mapsto 0$ or $f(x, y; t)dx \mapsto D_i(f)dx$, respectively, and compatibility with algebraic relations. Then one can check that the action goes down in a natural manner to $\mathcal{H}^1_{DR}(\mathcal{Y}/T)$ producing commutative diagrams

$$\begin{array}{ccc} \Omega^1_{\mathcal{Y}/T} & \xrightarrow{D_i} & \Omega^1_{\mathcal{Y}/T} \\ \downarrow & & \downarrow \\ \mathcal{H}^1_{DR}(\mathcal{Y}/T) & \xrightarrow[D_i]{} & \mathcal{H}^1_{DR}(\mathcal{Y}/T) \,. \end{array}$$

We see that the $\mathbb{C}[T]$-module $H^1_{DR}(\mathcal{Y}/T)$ has the following additional structure:

$$\begin{array}{lll} \text{(i)} & H^1_{DR}(\mathcal{Y}/T) \text{ is a } \mathbb{C}[T, D]\text{-module}, & (3.17) \\ \text{(ii)} & H^1_{DR}(\mathcal{Y}/T) \text{ is a } G\text{-module}, & \end{array}$$

where $\mathbb{C}[T, D] = \mathbb{C}[T][D_1, \ldots, D_r]$ is the ring of differential operators of T and $G = \mathbb{Z}/n\mathbb{Z}$ the cyclic GALOIS group of order n acting on $\mathcal{Y}$. With respect to the G-action (ii) we obtain an isotypical decomposition

$$H^1_{DR}(\mathcal{Y}/T) = \bigoplus_{\ell=0}^{n-1} H^1_{DR}(\mathcal{Y}/T)_\ell \tag{3.18}$$

in mod $\mathbb{C}[T, D]$ with dx/y^ℓ in $H^1_{DR}(\mathcal{Y}/T)_\ell$. We have similar decompositions for sheaves. With obvious notations we have an exact sequence of quasicoherent $\mathcal{O}_T$-module sheaves:

$$(\mathcal{O}_\mathcal{Y})_\ell \xrightarrow[d_{\mathcal{Y}/T}]{} (\Omega^1_{\mathcal{Y}/T})_\ell \longrightarrow \mathcal{H}^1_{DR}(\mathcal{Y}/T)_\ell \longrightarrow 0 \tag{3.19}$$

In order to find a minimal set of generators of $H^1_{DR}(\mathcal{Y}/T)$ over the non-commutative ring $\mathbb{C}[T, D]$ we organized in [33] an effective "bombardment" of the kernel of the middle map in (3.19) via $d = d_{\mathcal{Y}/T}$ with regard to the following steps:

1. Find a store of rational functions f on T such that
$$df = p(t, D)(dx/y^\ell) \,, \quad p(t; D) \in \mathbb{C}[T, D] \,.$$
Then, obviously, the hypercohomology class $\overline{dx/y^\ell}$ satisfies the differential equation
$$p(t, D)\overline{dx/y^\ell} = 0 \,.$$
We succeeded in finding several "cubic" and "quadratic" operators $p(t, D)$ with respect to the degree of p in $D_1, \ldots, D_r$.
2. Enlarge and clean the store by suitable combinations in order to remove as many D-quadratic summands as possible in the quadratic operators we found in step 1. not leaving our store.

The first structure result is the following:

Porposition 3.19.
$$H^1_{DR}(\mathcal{Y}/T)_\ell = \mathbb{C}[T, D]\overline{dx/y^k} \text{ for a suitable } k \equiv l \bmod n, k > 0 \,.$$

We assume from now that the curve family (3.12) is *primitive.* This means that g.c.d. $(n, r+2) = 1$. Also we exclude $l \equiv 0 \bmod n$.

Definition-Proposition 3.20. In the annulator $\mathrm{Ann}(dx/y^\ell)$ consisting of all differential operators $p \in \mathbb{C}[T, D]$ killing dx/y^ℓ there exists for each pair $1 \le i, j \le r$ one and only one quadratic operator ε_{ij} with $D_i D_j$ as the only D-quadratic summand.

The ε_{ij} with $i \neq j$ are called EULER *operators* and the ε_{ii} PICARD *operators.*

Both, EULER and PICARD operators are explicitly known, see [33]. Since the EULER operators are very simple we reproduce them here:

$$\varepsilon_{ij} = D_i D_j + \frac{\ell}{n(t_j - t_i)}(D_i - D_j) \,. \tag{3.20}$$

For the PICARD operators we refer within this book only to the special cases $\varepsilon_{11} = D_{11}$, $\varepsilon_{22} = D_{22}$ in Theorem 3.6.

Definition-Proposition 3.21. *With the above notations the* EULER-PICARD *system of partial differential equations in r variables*

$$\varepsilon_{ij} F = 0 \,, \quad 1 \le i \,, \, j \le r \tag{3.21}$$

is integrable and has locally $r+1$ basic solutions, this means that these solutions $F_0, \ldots, F_r$, say, are linearly independent and any other (local) solution is a $\mathbb{C}$-linear *combination of them.*

For suitable (local) cycle families $\underline{\alpha}_0, \ldots, \underline{\alpha}_r$ of $\tilde{C}/T$ the corresponding hypergeometric integrals

$$I_0 = \int\limits_{\underline{\alpha}_0} dx/y^\ell, \ldots, I_r = \int\limits_{\underline{\alpha}_r} dx/y^\ell \tag{3.22}$$

is a set of (local) basic solution of the EULER-PICARD system (3.21).

Remark 3.22 ([33]). The left ideal sheaf generated by the EULER-PICARD operators defined in Proposition-Definition 3.1 coincides with the left ideal sheaf of differential operators killing the hypergeometric integral functions $I_0, \ldots, I_r$ defined in (3.22).

Looking for a family with a section $\bar{\omega}$ in $\underline{\mathcal{H}}^1_{DR}(\tilde{C}/T)$ satisfying the differential equations (3.10) with $\bar{\omega}$ instead of F one can take the family of PICARD curves

$$\tilde{C}/(\mathbb{P}^2\backslash\triangle\!\!\!\!\star) = C/(\mathbb{P}^2\backslash\triangle\!\!\!\!\star) : Y^3 = X(X-1)(X-u)(X-v) \ ,$$

and $\bar{\omega}$ represented by the differential form $\omega = dx/y$ depending on u, v. Taking integrals along cycles one gets an "algebraic" fundamental system of solutions

$$I_k(t) = \int\limits_{\underline{\alpha}_k(t)} \omega(t) \ , \quad k = 0, 1, 2 \ , \ t = (u, v) \in \mathbb{P}^2\backslash\triangle\!\!\!\!\star) \ , \ \omega = dx/y \ . \tag{3.23}$$

Altogether we found the developing map of the Fuchsian system (3.10) in an explicit and algebraic manner. Looking back to the geometric starting point 3.7 and to the results of Proposition 3.13, Theorems 3.17 and 3.18 we obtain

Theorem 3.23. *The quotient map* $p : \mathbb{B} \to \mathbb{P}^2$ *with covering group* $\Gamma(\sqrt{-3})$ *is inverted by* $(I_0 : I_1 : I_2) : \mathbb{P}^2\backslash\triangle\!\!\!\!\star \dashrightarrow \mathbb{B}$ *on* $\mathbb{P}^2\backslash\triangle\!\!\!\!\star$ *with hypergeometric integrals* $I_k(t)$ *described in* (3.10) *along linearly independent cycle families* $\underline{\alpha}_0(t)$, $\underline{\alpha}_1(t)$, $\underline{\alpha}_2(t)$.

Remark 3.24. Historically not p but p^{-1} was roughly known first [60]. The monodromy group was known to be a sublattice of $\Gamma(\sqrt{-3})$ generated by five elements, see [1]. The more precise result of Theorem 3.23 together with Corollary 3.16 has been established in [33].

3.3 PICARD Modular Forms

We recall to TORELLI's Theorem 2.11 and its finer version in Theorem 2.34. For $g = 3$ we have an embedding $\mathcal{M}_3 \to \mathcal{A}_3$. It restricts to the moduli space $\mathcal{M}_3' \subset \mathcal{M}_3$ of smooth PICARD curves. We "uniformized" already the embedding $\mathcal{M}_3' = \mathbf{M}^0 \to \mathcal{A}_3$ in 2.7, see diagram (2.63) there. Together with the right-hand side of diagram

(2.59) precisely determined at the end of section 3.1, we can now establish the following SCHOTTKY-TORELLI *diagram* (3.24) for PICARD curves.

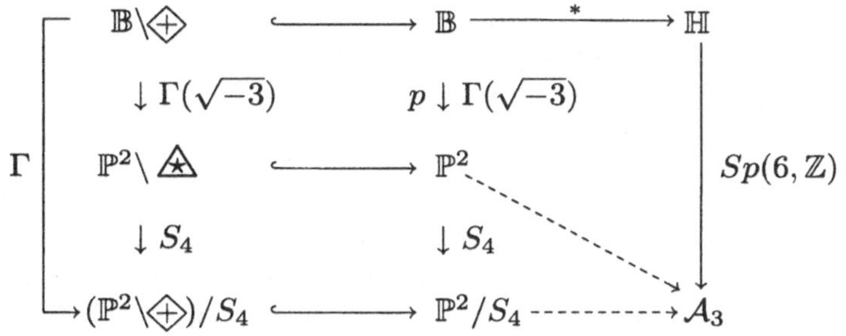

$$
\Gamma \left[
\begin{array}{ccccc}
\mathbb{B}\backslash \diamondsuit & \longleftrightarrow & \mathbb{B} & \xrightarrow{\;*\;} & \mathbb{H} \\
\downarrow \Gamma(\sqrt{-3}) & & p \downarrow \Gamma(\sqrt{-3}) & & \Big\downarrow \\
\mathbb{P}^2\backslash \triangle & \longleftrightarrow & \mathbb{P}^2 & & Sp(6,\mathbb{Z}) \\
\downarrow S_4 & & \downarrow S_4 & & \\
(\mathbb{P}^2\backslash\diamondsuit)/S_4 & \longleftrightarrow & \mathbb{P}^2/S_4 & \dashrightarrow & \mathcal{A}_3
\end{array}
\right] \qquad (3.24)
$$

where the broken lines denote rational maps.

By TORELLI, the isomorphy class of a smooth curve is uniquely determined by its (polarized) Jacobian variety. We look for a precise pointwise version of this theorem in the case of PICARD curves. The corresponding "effective TORELLI problem", see 2.12.4, goes down along the upper part of diagram (2.63) to the following reformulation:

3.25. Find for given $\tau \in \mathbb{B}$ (or $*\tau \in *\mathbb{B} \subset \mathbb{H}_3$) the normal form of a PICARD curve C_τ corresponding to the moduli point $p(\tau) \in \mathbb{P}^2$.

In analogy to the WEIERSTRASS normal form of elliptic curves we are able to prove

Proposition 3.26. *There are holomorphic functions* t_1, t_2, t_3, t_4 *on* $\mathbb{B}$ *such that the normal forms we look for can be written as*

$$
C_\tau : Y^3 = \prod_{i=1}^{4} (X - t_i(\tau)) \qquad (3.25)
$$

simultaneously for all $\tau \in \mathbb{B}$.

In other words, we try to describe the quotient map p in terms of holomorphic functions identifying the quotient map p with

$$
(t_1 : t_2 : t_3 : t_4) : \mathbb{B} \longrightarrow \mathbb{P}^2 \qquad (3.26)
$$
$$
\tau \longmapsto (t_1(\tau) : t_2(\tau) : t_3(\tau) : t_4(\tau)) \ .
$$

The presentation of these holomorphic functions was the main result of chapter I in [33]. We refer the reader to section 6.3 there, entitled "Inversion of the PICARD integral map by means of automorphic forms", especially to Theorem 6.3.12.

Since we need the quality of the functions t_i for finding explicit FOURIER series of them, we repeat the way of their construction in [33] without proofs.

Definition 3.27. A holomorphic function $f : \mathbb{B} \to \mathbb{C}$ is a PICARD *modular form* of the imaginary quadratic number field K and of weight m, if there exists a sublattice Γ'' of $\Gamma_K = \mathbb{U}((2,1), \mathcal{O}_K)$ such that the following *functional equations* are satisfied:

$$\gamma^*(f) = j_\gamma^m \cdot f \quad \text{for all} \quad \gamma \in \Gamma'' , \tag{3.27}$$

where $\gamma^*(f)(\tau) = f(\gamma(\tau))$ and $j_\gamma(\tau)$ is the JACOBI determinant of $\gamma : \mathbb{B} \xrightarrow{\sim} \mathbb{B}$ at τ.

If (3.27) is satisfied, then we shortly call f a Γ''-*modular form* (of *weight* m). These functions form a finite-dimensional vector space denoted by $[\Gamma'', m]$. We come back now to the EISENSTEIN numbers, especially to $\Gamma = \mathbb{U}((2,1), \mathcal{O}_K)$, $\Gamma' = \Gamma(\sqrt{-3})$ and define *special* PICARD *modular groups* by

$$\mathbb{S}\Gamma = \Gamma \cap \mathbb{S}l_3(\mathbb{C}) , \quad \mathbb{S}\Gamma(\sqrt{-3}) = \Gamma(\sqrt{-3}) \cap \mathbb{S}l_3(\mathbb{C}) .$$

We have the exact commutative diagram

$$\begin{array}{ccccccccc} & & 1 & & 1 & & 1 & & \\ & & \downarrow & & \downarrow & & \downarrow & & \\ 1 & \longrightarrow & \mathbb{S}\Gamma(\sqrt{-3}) & \longrightarrow & \mathbb{S}\Gamma & \longrightarrow & S_4 & \longrightarrow & 1 \\ & & \downarrow & & \downarrow & & \downarrow & & \\ 1 & \longrightarrow & \Gamma(\sqrt{-3}) & \longrightarrow & \Gamma & \longrightarrow & S_4 \times Z_2 & \longrightarrow & 1 \\ & & \downarrow & & \downarrow \text{det} & & \downarrow & & \\ 1 & \longrightarrow & Z_3 & \longrightarrow & Z_6 & \longrightarrow & Z_2 & \longrightarrow & 1 \\ & & \downarrow & & \downarrow & & \downarrow & & \\ & & 1 & & 1 & & 1 & & \end{array} \tag{3.28}$$

For example, the group $Z_2 = \mathbb{Z}/2\mathbb{Z}$ comes from the element $-id \in \Gamma$.

In Proposition 3.26 we look for PICARD modular forms t_1, t_2, t_3, t_4 of small weight. Excluding constants the weight must be positive. So the weight should be 1 and the level group should be $\Gamma(\sqrt{-3})$. Unfortunately such non-trivial PICARD modular forms do not exist, see [33]. Our level group is a little bit too big. But the change to $\mathbb{S}\Gamma(\sqrt{-3})$ is successful. The factor group $\Gamma/\mathbb{S}\Gamma(\sqrt{-3}) = S_4 \times Z_6$ acts on the vector space $[\mathbb{S}\Gamma(\sqrt{-3}), 1]$ of $\mathbb{S}\Gamma(\sqrt{-3})$-forms of weight 1. By some geometric considerations and elementary representation theory one can hope to find a set of four special forms in $[\mathbb{S}\Gamma(\sqrt{-3}), 1]$ such that the elements of S_4 act as permutations on this set, up to a character. What we are going to realize is to find PICARD modular forms t_1, t_2, t_3, t_4 satisfying the following conditions (3.29) and 3.28:

$$t_1 + t_2 + t_3 + t_4 = 0 , \qquad t_1, t_2, t_3 \text{ are linearly independent.} \tag{3.29}$$

3.28 Special Functional Equations

(i) $\gamma^*(t_i) = j_\gamma \cdot t_i$ for $i = 1,2,3,4$, $\gamma \in \mathbb{S}\Gamma(\sqrt{-3})$;

(ii) $\gamma^*(t_i/t_j) = t_{\bar{\gamma}(i)}/t_{\bar{\gamma}(j)}$ for $i = 1,2,3,4$, $\gamma \in \mathbb{S}\Gamma$ with image $\bar{\gamma}$ in $S_4 = \mathbb{S}\Gamma/\mathbb{S}\Gamma(\sqrt{-3})$;

(iii) $\delta^*(t_i) = (\det\delta)^2 \cdot j_\delta \cdot t_i$ for $i = 1,2,3,4$, δ representing a generator of $\Gamma(\sqrt{-3})/\mathbb{S}\Gamma(\sqrt{-3})$.

Theorem 3.29 ([33], [35]). *There exist four* PICARD *modular forms* t_1, t_2, t_3, t_4 *satisfying the properties* (3.29) *and* 3.28. *They are uniquely determined up to the numeration and a common constant factor. The condition (iii) is a consequence of all the previous conditions* (3.29), 3.28 *(i), (ii).*

Remark 3.30. The last statement has been proved first by FEUSTEL [23] by an analytic argument using the Theta presentation of the modular forms t_i we look for in the next section. An algebraic-geometric proof was given by the author [35] by means of a dimension formula for cusp forms of ball lattices. The splitting of the special functional equations into three conditions is useful in the next section. But it can be written as one *Nebentypus* condition. For this purpose we denote the image of γ along $\Gamma \to S_4 \times Z_2 \to S_4$ (see diagram (3.28) by $\bar{\gamma}$. Then we get

Corollary 3.31. *The four* PICARD *modular forms* t_1, t_2, t_3, t_4 *we look for are characterized (up to a common constant factor) by* (3.29) *and the functional equations*

$$\gamma^*(t_i) = (\det\gamma)^2 \cdot sgn(\bar{\gamma}) \cdot j_\gamma \cdot t_{\bar{\gamma}(i)} \quad \text{for } i = 1,2,3,4\,.$$

Main idea of the proof (see [33]). Basically one has to classify the surface $\hat{X} = \widehat{\mathbb{B}/\mathbb{S}\Gamma}(\sqrt{-3})$. The group $\mathbb{S}\Gamma(\sqrt{-3})$ acts almost freely on $\mathbb{B}$, this means that the non-trivially acting elements have at most isolated fixed points on $\mathbb{B}$. It turns out that

$$\widehat{\mathbb{B}/\mathbb{S}\Gamma}(\sqrt{-3}) = \hat{X} \longrightarrow \mathbb{P}^2 = \widehat{\mathbb{B}/\Gamma}(\sqrt{-3})$$

is the unique 3-cyclic covering of $\mathbb{P}^2$ branched along $\triangle\!\!\!\!\star$ (see diagram (3.28) and Proposition 3.13). It is not difficult to describe the surface $\hat{X}$ by an equation (see [33], I.4.3):

$$\hat{X} : Z^3 = (Y_2^2 - Y_1^2)(Y_2^2 - Y_0^2)(Y_1^2 - Y_0^2) \tag{3.30}$$

This is a weighted equation with Z of weight 2 and Y_0, Y_1, Y_2 of weight 1. More precisely, this means that $\hat{X}$ is the projective spectrum of the corresponding graded ring

$$\mathbb{C}[Y_0, Y_1, Y_2, Z] \Big/ \left(Z^3 - \prod_{0 \le i < j \le 2} (Y_j^2 - Y_i^2) \right).$$

Remark 3.32. At this point we remind again to the 22nd HILBERT Problem. It turns out that the algebraic relation in (3.30) is satisfied by PICARD modular forms y_0, y_1, y_2, z substituting the variables Y_0, Y_1, Y_2 or Z, respectively. The knowledge of the uniformization $\mathbb{B} \to \hat{X}$ together with the corresponding arithmetic uniformizing lattice $\mathbb{S}\Gamma(\sqrt{-3})$ becomes important for this purpose as it has been predicted by HILBERT in a more general setting. A general solution has been given by the compactification theory via projective spectra of rings of automorphic forms, due to BAILY-BOREL [2]. Additionally, one has of GRIFFITHS' general Uniformization Theorem [25]: Each projective variety is the compactification of a quotient variety $\mathbb{V}/\Sigma$, $\mathbb{V}$ a bounded domain, Σ a discrete sublattice of the group $\mathrm{Aut}_{hol}(\mathbb{V})$ of biholomorphic automorphisms of $\mathbb{V}$. The theorems of BAILY-BOREL and GRIFFITHS are welcome guides. But an effective presentation of $\mathbb{V}$, Σ and the corresponding automorphic forms in special situations is not superfluous and remains difficult.

The key point is to understand automorphic forms as sections of logarithmic pluricanonical sheaves. In [33], I.4.3 we proved

$$\bigoplus_{m=0}^{\infty} [\mathbb{S}\Gamma(\sqrt{-3}), m] = \bigoplus_{m=0}^{\infty} H^0\left(\bar{X}, \mathcal{O}(mK_{\bar{X}} + mT)\right) , \tag{3.31}$$

where $\bar{X}$ is the minimal resolution of singularities of $\hat{X}$, T the compactification divisor resolving the cusp singularities of $\hat{X}$. It consists of four disjoint elliptic curves. As usual $K_{\bar{X}}$ denotes a canonical divisor and $\mathcal{O}(D)$ is the sheaf corresponding to the divisor D. A careful geometric analysis (explicit knowledge of a canonical divisor, vanishing theorem on surfaces) accomplished in [33] yields the ring structure

$$\bigoplus_{m=0}^{\infty} H^0\left(\bar{X}, \mathcal{O}(mK_{\bar{X}} + mT)\right) = \mathbb{C}[s_0, s_1, s_2, s] \tag{3.32}$$

with s_0, s_1, s_2 of weight 1, s of weight 2, and the generating relation

$$s^3 = (s_2^2 - s_1^2)(s_2^2 - s_0^2)(s_1^2 - s_0^2) . \tag{3.33}$$

Together with (3.31) we found generators η_0, η_1, η_2 of $[\mathbb{S}\Gamma(\sqrt{-3}), 1]$ and an $\mathbb{S}\Gamma(\sqrt{-3})$-modular form η of weight 2 such that η_0, η_1, η_2 and η generate the ring $\bigoplus_{m=0}^{\infty} [\mathbb{S}\Gamma(\sqrt{-3}), m]$ of $\mathbb{S}\Gamma(\sqrt{-3})$-modular forms satisfying the relation

$$\eta^3 = (\eta_2^2 - \eta_1^2)(\eta_2^2 - \eta_0^2)(\eta_1^2 - \eta_0^2) \tag{3.34}$$

we were looking for.

If Γ'' is an arbitrary ball lattice, then it acts on the space of holomorphic functions on the ball $\mathbb{B}$ via

$$f \mapsto j_\gamma^{-m} \cdot \gamma^*(f) , \quad f \in H^0(\mathbb{B}, \mathcal{O}_{\mathbb{B}}) , \quad \gamma \in \Gamma'' . \tag{3.35}$$

The Γ''-invariant functions are the Γ''-modular forms (compare with (3.27)). In particular, the lattice $\mathbb{S}\Gamma$ acts on $[\mathbb{S}\Gamma(\sqrt{-3}),1]$ with ineffective kernel $\mathbb{S}\Gamma(\sqrt{-3})$. With regard to diagram (3.28) we get a three-dimensional representation of S_4. In [33] we proved that this representation is irreducible. It induces a projective representation of S_4 on $\mathbb{P}[\mathbb{S}\Gamma(\sqrt{-3}),1] \cong \mathbb{P}^2$. There is only one such representation, up to equivalence. Explicitly it can be described by

$$(x_1 : x_2 : x_3 : x_4) \mapsto \left(x_{\sigma(1)} : x_{\sigma(2)} : x_{\sigma(3)} : x_{\sigma(4)}\right)\ , \quad \sigma \in S_4\,,\ x_i \in \mathbb{C}\,,\ \Sigma x_i = 0\,.$$

Looking back to $[\mathbb{S}\Gamma(\sqrt{-3}),1]$ one finds four PICARD modular forms t_1, t_2, t_3, t_4 satisfying (3.29) and 3.28 (i), (ii).

It remains to verify the property (iii) of 3.28. This is much more difficult than it looks at first glance, comparable with some "sign questions" in number theory. As we already remarked in Remark 3.30 there exist two proofs of different kind. For a "transcendental proof" via theta constants we refer also to SHIGA's articles [74].

In order to solve our relative TORELLI problem in an effective manner by means of the modular forms t_1, t_2, t_3, t_4 we go back to the quotient map (3.26). It is realized by the modular forms of Theorem 3.29 for the following reasons (see [33] for more details). From the left column of (3.28) one gets a commutative quotient diagram

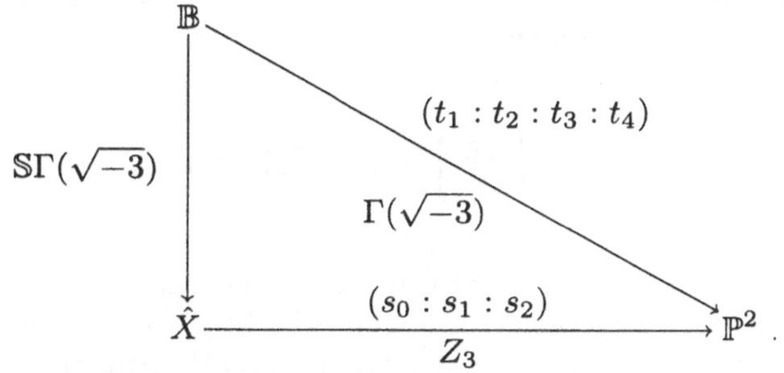

(3.36)

The logarithmic canonical map $\varphi_{K_{\bar{X}}+T}$ goes down to $\hat{X} \longrightarrow \mathbb{P}^2$ and coincides with the Z_3-quotient map on the bottom of diagram (3.36) as has been verified in [33]. Using generators s_i of $H^0\left(\bar{X}, \mathcal{O}(K_{\bar{X}} + T)\right)$ it can be realized as the projective morphism $(s_0 : s_1 : s_2)$. The sections s_i have been lifted to $\mathbb{S}\Gamma(\sqrt{-3})$-modular forms η_i, $i = 0,1,2$ via (3.31). Without loss of generality we can assume that $t_i = \eta_{i+1}$. Then the $\Gamma(\sqrt{-3})$–quotient map of (3.36) coincides with the analytic map in (3.26), where t_4 is defined by (3.29).

Theorem 3.33 (effective TORELLI theorem for PICARD curves in terms of PICARD modular forms). *Let $Jac(C)$ be the polarized Jacobi threefold of a smooth* PICARD *curve C corresponding to the point $\Omega = *\tau \in \mathbb{H}_3$. Then the* PICARD *modular forms t_1, t_2, t_3, t_4 of Theorem* 3.29 *recover a normal form of C in the following manner:*

$$C \cong C' : Y^3 = (X - t_1(\tau))(X - t_2(\tau))(X - t_3(\tau))(X - t_4(\tau))\,. \tag{3.37}$$

Remember first that we solved already the relative SCHOTTKY problem for PICARD curves, see 2.12, the end of 2.7 (rough solution), the end of 3.1 (fine solution), the rough diagram (2.63) and the SCHOTTKY-TORELLI diagram (3.24). Altogether we proved the following

Theorem 3.34 (Relative SCHOTTKY for PICARD curves). *The matrix* Π *in* (2.28) *is a period matrix of a smooth* PICARD *curve if and only if the following condition is satisfied:*

$$\Pi \in \mathbb{G}l_3(\mathbb{C}) \begin{pmatrix} *\mathfrak{a} \\ \hline *\mathfrak{b} \\ \hline *\mathfrak{e} \end{pmatrix} \mathbb{S}p(6, \mathbb{Z})$$

where the $*$*-matrix is typical, that is*

(i) $\mathfrak{a} \in \mathbb{C}^3$, $\langle \mathfrak{a}, \mathfrak{a} \rangle < 0$ *(ball condition);*

(ii) $\mathfrak{b}, \mathfrak{e}$ *is a basis of* $\mathfrak{a}^\perp$ *(orthogonal condition);*

(iii) $\tau = \mathbb{P}\mathfrak{a}$ *does not belong to* $\bigoplus$ *defined in* (3.8) *(regular condition).* □

Proof of Theorem 3.33: By the Relative SCHOTTKY Theorem 3.34 for PICARD curves we have $J(C) = \mathbb{C}^3/\Lambda_\Pi$, where Π is assumed to be typical. We have a unique composed correspondence

$$* : \tau = \mathbb{P}\mathfrak{a} \mapsto \begin{pmatrix} *\mathfrak{a} \\ \hline *\mathfrak{b} \\ \hline *\mathfrak{e} \end{pmatrix} = (\Pi_1|\Pi_2) \underset{\Pi_1^{-1}}{\longmapsto} (E_3|\Omega) \mapsto \Omega \tag{3.38}$$

connecting $\tau \in \mathbb{B}$ with $\Omega = *\tau$. The SCHOTTKY-TORELLI diagram (3.24) yields the moduli point of C on $\mathbb{P}^2$ as image of τ along the $\Gamma(\sqrt{-3})$-quotient map. By (3.36) this image is equal to $(t_1(\tau) : t_2(\tau) : t_3(\tau) : t_4(\tau))$. But the normal form of a PICARD curve of this moduli point is given by the equation in (3.37), see 2.1. For the convenience of the reader we present a diagram of correspondences used above in close connection with diagram (3.24).

$$\begin{array}{ccc} \mathbb{B} \ni \tau = \mathbb{P}\mathfrak{a} & \longleftarrow & \Pi = \begin{pmatrix} *\mathfrak{a} \\ \hline *\mathfrak{b} \\ \hline *\mathfrak{c} \end{pmatrix} \\ \big\downarrow & & \big\updownarrow \\ C : Y^3 = \prod_{i=1}^{4} (X - t_i(\tau)) & \longleftarrow\!\!\text{-}\,\text{-}\,\text{-} & \mathrm{Jac}\,(C) \end{array}$$

where broken arrows represent multivalued maps. Theorem 3.33 is proved. □

3.4 PICARD Modular Forms as Theta Constants

Theta functions $\vartheta \left[\begin{smallmatrix} a \\ b \end{smallmatrix}\right]$ with characteristics $a, b \in \mathbb{Q}^g$ are holomorphic functions on $\mathbb{C}^g \times \mathbb{H}_g$, $\mathbb{H}_g$ the generalized SIEGEL upper half plane uniformizing the moduli space of (principally) polarized abelian varieties of dimension g. Explicitly the theta functions

$$\vartheta \begin{bmatrix} a \\ b \end{bmatrix} : \mathbb{C}^g \times \mathbb{H}_g \longrightarrow \mathbb{C}$$

are defined by

$$\vartheta \begin{bmatrix} a \\ b \end{bmatrix} (z, \Omega) = \sum_{n \in \mathbb{Z}} \exp \left\{ \pi i^t (n + a) \Omega (n + a) + 2\pi i^t (n + a)(z + b) \right\} .$$

The restrictions $\vartheta \mid 0 \times \mathbb{H}_g$,

$$\theta \begin{bmatrix} a \\ b \end{bmatrix} (\Omega) = \vartheta \begin{bmatrix} a \\ b \end{bmatrix} (0, \Omega)$$

are called *theta constants* (with *characteristics* a, b).

We restrict our attention to the case $g = 3$ and look for extensions of the PICARD modular forms t_1, t_2, t_3, t_4, defined in the previous section, on $\mathbb{H}_3$ along the embedding $* : \mathbb{B} \hookrightarrow \mathbb{H}_3$, see (3.24), and hope to be able to express them in terms of theta constants. Very important for this purpose are the functional equations described in 3.28. We look for elementary combinations Th of theta constants whose restrictions

$$th(\tau) = Th(*\tau) , \quad \tau \in \mathbb{B}$$

satisfy the special functional equations 3.28 (i), (iii):

$$Th \circ (*\gamma) = (\det \gamma)^2 \cdot j_\gamma \cdot Th \text{ on } \mathbb{B} \subset \mathbb{H}_3 , \ \gamma \in \Gamma(\sqrt{-3}) . \tag{3.39}$$

For the convenience of the reader we summarize the restrictions (or extensions) used above and below in the following diagram:

$$\begin{array}{rcccl} & \mathbb{B} & \hookrightarrow & \mathbb{H} & \\ \Gamma(\sqrt{-3}) \ni \gamma \downarrow\wr & & & \downarrow\wr * \gamma \in \mathbb{S}p(6, \mathbb{Z}) & \\ & \mathbb{B} & \hookrightarrow & \mathbb{H}_3 & \hookrightarrow \mathbb{C}^3 \times \mathbb{H}_3 \\ th = \theta|_{\mathbb{B}} \downarrow & & & \downarrow \theta = \vartheta|_{O \times \mathbb{H}_3} & \quad \downarrow \vartheta \\ & \mathbb{C} & \xrightarrow{=} & \mathbb{C} & \xrightarrow{\quad = \quad} \mathbb{C} \end{array} \tag{3.40}$$

Theorem 3.35 (FEUSTEL, SHIGA). *Let* $\theta_i(\Omega) = \vartheta_i(0, \Omega)$, $i = 0, 1, 2$, *be the theta constants on* $\mathbb{H}_3$ *restricting the theta functions*

$$\vartheta_k(z, \Omega) = \vartheta \begin{bmatrix} 0 & 1/6 & 0 \\ k/3 & 1/6 & k/3 \end{bmatrix} (z, \Omega) , \quad k = 0, 1, 2 , \ z \in \mathbb{C}^3 .$$

Set

$$Th_1 = \theta_0^3 + \theta_1^3 + \theta_2^3, \quad Th_2 = -3\theta_0^3 + \theta_1^3 + \theta_2^3 , \tag{3.41}$$
$$Th_3 = \theta_0^3 - 3\theta_1^3 + \theta_2^3, \quad Th_4 = \theta_0^3 + \theta_1^3 - 3\theta_2^3 ,$$

and

$$th_i(\tau) = Th_i(*\tau) , \quad i = 1, 2, 3, 4 , \ \tau \in \mathbb{B} . \tag{3.42}$$

Then the functions $th_i(\tau)$ *are the normalized* PICARD *modular forms satisfying* (3.29) *and all the functional equations (i), (ii), (iii) of* 3.28 *or, equivalently, those of Corollary* 3.31.

Proof (main steps): We follow FEUSTEL's proof and refer for explicit calculations to his paper [23] and the related literature given there. The proof summarizes preparatory work of RIEMANN, PICARD [60], [61], ALEZAIS [1], MUMFORD [56], H. SHIGA [74] and HOLZAPFEL [33].
Step 1: Restriction to six functional equations.

Here we go back to the fundamental group $\pi_1(\mathbb{P}^2\backslash\bigstar)$ of the Fuchsian system (3.10) of partial differential equations and the surjective monodromy representation $\pi_1(\mathbb{R}^2\backslash\bigstar) \to \Gamma(\sqrt{-3})$, coming from the linear action of the fundamental group on the three-dimensional space of solutions of the system, see Remark 3.24 and [33]. But the fundamental group $\pi_1(\mathbb{P}^2\backslash\bigstar)$ has obviously six generators coming from simple loops in $\mathbb{P}^2$ around each one of the six omitted lines. Therefore also $\Gamma(\sqrt{-3})$ has six generators, say $g_1, \ldots, g_6$. They have been explicitly described already by PICARD [60] (with correction in [61]) and ALEZAIS. Their symplectic lifts $G_i = *g_i \in \mathbb{S}p(6, \mathbb{Z})$, $i = 1, \ldots, 6$, can be found explicitly by means of relation (2.50). In order to check the functional equations 3.28 (i), (iii) for suitable holomorphic functions th on $\mathbb{B}$ it is sufficient to check them for the generators of $\Gamma(\sqrt{-3})$. According to our claim $th = Th|_{\mathbb{B}}$ we have now only to look for holomorphic functions Th on $\mathbb{H}_3$ satisfying the six *restricted functional equations*

$$Th \circ G_i = (\det g_i)^2 \cdot j_{g_i} \cdot Th \text{ on } \mathbb{B} \subset \mathbb{H}_3 , \quad i = 1, \ldots, 6. \tag{3.43}$$

Step 2: RIEMANN's Theorem.

It is a general problem in the theory of algebraic curves to describe a given meromorphic function on a curve C in terms of theta functions on its Jacobian variety by restriction along the JACOBI embedding $C \to J(C)$. This problem has been solved essentially by RIEMANN. We refer to MUMFORD's book [56].

Theorem 3.36 (RIEMANN). *Let C be a (smooth, compact, complex) curve of positive genus g, $\alpha_1, \dots, \alpha_{2g}$ a normal basis of $H_1(C, \mathbb{Z})$ and $\vec{\omega} = (\omega_1, \dots, \omega_g)$ a basis of $H^0(C, \omega_C)$ such that the corresponding period matrix has the normalized form*

$$\left(\int_{\alpha_j} \omega_i \right) = (E_g \mid \Omega) , \quad \Omega \in \mathbb{H}_g , \quad E_g \text{ the unit matrix.}$$

If $f : C \to \mathbb{P}^1$ is a meromorphic function with divisor

$$(f) = \sum_{k=1}^{m} a_k - \sum_{k=1}^{m} b_k , \quad a_k, b_k \in C ,$$

then it holds that

$$\prod_{i=1}^{g} f(P_i) = \text{const} \cdot \prod_{k=1}^{m} \left\{ \vartheta \left(\sum_{i=1}^{g} \int_{P_0}^{P_i} \vec{\omega} - \int_{P_0}^{a_k} \vec{\omega} - \Delta \right) / \vartheta \left(\sum_{i=1}^{g} \int_{P_0}^{P_i} \vec{\omega} - \int_{P_0}^{b_k} \vec{\omega} - \Delta \right) \right\} \tag{3.44}$$

as meromorphic function on C^g/S_g, where one has to use the same paths in the first integrals of the denominator and numerator.

Notations. Here the RIEMANN theta function ϑ is considered as holomorphic function on $\mathbb{C}^g$. It coincides with the restriction of $\vartheta \left[\begin{smallmatrix} 0 \\ 0 \end{smallmatrix} \right]$ to $\mathbb{C}^g \times \Omega$, Ω the fixed period matrix defined above, with the notations introduced at the beginning of this section. The auxiliary point $P_0 \in C$ is used to fix the JACOBI embedding

$$C \to J(C) , \ P \mapsto \int_{P_0}^{P} \vec{\omega} \bmod \Lambda_\Omega , \quad \Lambda_\Omega = \mathbb{Z}^g + \Omega \mathbb{Z}^g .$$

Δ denotes the RIEMANN constant. This is a special well-defined 2-torsion point on $J(C)$ (see [56], ch. II,3). Both sides are considered as functions on C^g or C^g/S_g, namely $\sum_{i=1}^{g} P_i$ is understood as point on C^g/S_g. On the right-hand side of (3.44) appears a constant denoted by const. In general one knows only its existence but not its explicit value.

Remark. For the proof of RIEMANN's Theorem one compares zeros and poles of both sides of (3.44). A key point is to understand the RIEMANN constant Δ. On $\mathbb{C}^g$ it is defined mod Λ_Ω by

$$-\Delta + \sum_{i=1}^{g-1} \int_{P_0}^{P_i} \vec{\omega} \bmod \Lambda_\Omega \ \in \ \Theta, \qquad P_i \in C ,$$

where $\Theta \subset J(C)$ denotes the theta divisor defined by $\vartheta(z) = 0$. Setting e.g. $P_i = a_k$ in (3.44) opens the way of proof of RIEMANN's Theorem by comparison of zeros and poles.

Now we apply RIEMANN's Theorem for finding two generators of the field of $\Gamma(\sqrt{-3})$-automorphic functions in theta terms. This has been done already by PICARD and ALEZAIS. We write a smooth PICARD curve C in the modified normal form

$$C : Y^3 = X(X-1)(X-u)(X-v) .$$

Then $u = u(\tau)$, $v = v(\tau)$, $\tau \in \mathbb{B}$, generate the field of $\Gamma(\sqrt{-3})$-modular functions. The ramification locus of the Z_3-GALOIS covering $C \to \mathbb{P}^1$ consists of the following five points on C:

$$0 = Q_0 = (0,0) \ , \ Q_1 = (1,0) \ , \ Q_u = (u,0) \ , \ Q_v = (v,0), \ \infty .$$

We apply (3.44) to the function $f = x : C \to \mathbb{P}^1$ at the points

$$P_1 = Q_1 \, , \ P_2 = Q_1 \, , \ P_3 = Q_u$$

and at the points

$$P_1 = Q_1 \, , \ P_2 = Q_u \, , \ P_3 = Q_u \, .$$

With $(x) = 3 \cdot O - 3 \cdot \infty$, $P_0 = \infty$, the same constant c and the general abbreviation

$$\int_{P_0}^{D} \vec{\omega} = \sum_{i=1}^{g} \int_{P_0}^{P_i} \vec{\omega} \quad \text{for } D = \sum_{i=1}^{g} P_i$$

one gets

$$u = c \cdot \prod_1^3 \left\{ \vartheta \left(\int_{\infty}^{2Q_1+Q_u} \vec{\omega} - \int_{\infty}^{0} \vec{\omega} - \Delta \right) / \vartheta \left(\int_{\infty}^{2Q_1+Q_u} \vec{\omega} - \Delta \right) \right\} ,$$

$$u^2 = c \cdot \prod_1^3 \left\{ \vartheta \left(\int_{\infty}^{Q_1+2Q_u} \vec{\omega} - \int_{\infty}^{0} \vec{\omega} - \Delta \right) / \vartheta \left(\int_{\infty}^{Q_1+2Q_u} \vec{\omega} - \Delta \right) \right\} .$$

Since $u = u^2/u$ we get by division of both expressions above a fractional theta formula for $u(\tau)$ without the unknown constant c. In the same manner, using Q_v instead of Q_u we can express v, v^2 and finally $v(\tau) = v^2/v$ in terms of the theta function belonging to $\Omega = *\tau$.

Step 3: Theta constants.

This step is due to SHIGA [74]. He calculated explicitly the RIEMANN constant Δ above using special values. Furthermore he used standard transformation laws for theta functions to prove that

$$u(\tau) = \vartheta_1^3(0,*\tau)/\vartheta_2^3(0,*\tau)\ , \quad v(\tau) = \vartheta_0^3(0,*\tau)/\vartheta_2^3(0,*\tau) \tag{3.45}$$

with the notations of Theorem 3.35. The denominator does not vanish identically on $\mathbb{B}$ (SHIGA [74], FEUSTEL [23]).

Step 4: Automorphic forms.

This final step is due to FEUSTEL [23]. He checked that the denominator and the numerators in (3.45) satisfy the six functional equations (3.43). Then we have on three linearly independent $\mathbb{S}\Gamma(\sqrt{-3})$-modular forms $\theta_0^3(*\tau)$, $\theta_1^3(*\tau)$, $\theta_2^3(*\tau)$ on $\mathbb{B}$ of weight 1 with the notations of Theorem 3.35. Since $\dim\left[\mathbb{S}\Gamma(\sqrt{-3}),1\right] = 3$ by the considerations around (3.33) we find by linear combinations all $\mathbb{S}\Gamma(\sqrt{-3})$-modular forms of weight 1. So $\theta_0^3, \theta_1^3, \theta_2^3|_{\mathbb{B}}$ can be identified with η_0, η_1, η_2 in (3.34). Now we refer to the $S_4 = \mathbb{S}\Gamma/\mathbb{S}\Gamma(\sqrt{-3})$-action on $[\mathbb{S}\Gamma(\sqrt{-3}),1]$. There must be linear combinations th_1, th_2, th_3, th_4 of $\theta_0^3, \theta_1^3, \theta_2^3|_{\mathbb{B}}$ satisfying relation (3.29) and the functional equations 3.28, uniquely defined up to a common factor, see Theorem 3.29. The symmetric group S_4 is generated by three transpositions. It is not difficult to find representations of them in $\mathbb{S}\Gamma$ and also their symplectic representations acting on $\mathbb{H}_3$, explicitly. This has been done in [23]. With the definitions (3.41) and (3.42) one can check that th_1, th_2, th_3, th_4 are the functions with the correct transformation behaviour we are looking for. Theorem 3.35 is proved. □

4 Algebraic Values of Picard Modular Theta Functions

4.1 Introduction

We define a *solution model* of the 12th HILBERT Problem to be an **explicit** triple (V, V_{sing}, f) consisting of an open (analytic) subspace V of a complex variety $W \subseteq \mathbb{P}^N$, a dense subset V_{sing} of $V(\overline{\mathbb{Q}}) = V \cap \mathbb{P}^N(\bar{\mathbb{Q}})$, and a transcendental holomorphic map

$$f = (f_0 : f_1 : \ldots : f_n) : V \to \mathbb{P}^n, \quad P \mapsto (f_0(P) : \ldots : f_n(P))$$

if the following holds:

0. $\dim V > 1$

I. For $\sigma \in V(\bar{\mathbb{Q}})$ the point $f(\sigma)$ is algebraic (i.e. $f(\sigma) \in \mathbb{P}^n(\bar{\mathbb{Q}})$) iff $\sigma \in V_{sing}$.

II. For $\sigma \in V_{sing}$ there is a purely number theoretic construction or description of field extensions $k(\sigma)(f(\sigma))/k(\sigma)/\mathbb{Q}$, where the basic ("reflex") field $k(\sigma)$ is an elementary substitute of the coordinate field $\mathbb{Q}(\sigma)$.

We call f *transcendental*, if it is not the restriction of an algebraic map.

A *ball model* for HILBERT's 12th Problem is a solution model $(\mathbb{B}, \mathbb{B}_{sing}, f)$, where $\mathbb{B} \subset \mathbb{P}^2$ is projectively isomorphic to the standard 2-ball

$$\mathbb{B}^2 = \left\{(z_1, z_2) \in \mathbb{C}^2 , \ |z_1|^2 + |z_2|^2 < 1\right\}.$$

The main purpose of this chapter is to understand the PICARD modular theta map $(th_1 : th_2 : th_3 : th_4) : \mathbb{B} \to \mathbb{P}^2$ constructed in chapter 3 as a ball model of HILBERT's 12th Problem in the above sense.

Historically, the origin of the the 12th Problem is KRONECKER's work on the explicit description of abelian extensions over the field $\mathbb{Q}$ of rational numbers or over imaginary quadratic number fields, respectively, by means of special values of special transcendent functions of one complex variable.

The Theorem of KRONECKER-WEBER asserts that each absolute abelian number field is generated over $\mathbb{Q}$ by a rational expression of a unit root, where unit roots are understood as special values of the exponential function.

As a counter part appears HILBERT's 7-th Problem. It asks for the quality of values of the shifted exponential function

$$e(z) = \exp(\pi i z) , \quad i = \sqrt{-1} ,$$

at algebraic arguments outside of the rationals $\mathbb{Q}$ and conjectures that all these values are transcendental numbers (see [30]). This problem has been solved affirmatively and independently by GELFOND [24] and SCHNEIDER [65] in 1934. Altogether we know:

I. $e(z)$ has algebraic values on $\mathbb{Q}$;

I'. $e(z)$ has transcendental values on $\bar{\mathbb{Q}}\backslash\mathbb{Q}$;

II. the number theoretic meaning of the values $e(q)$, $q \in \mathbb{Q}$.

Substituting the base field $\mathbb{Q}$ by an imaginary quadratic number field one needs special values of WEIERSTRASS' $\wp$-function (at torsion points of an elliptic curve), and special values (at singular moduli) of the elliptic modular function j in order to generate all abelian extensions. For a precise formulation of this classical Main Theorem of Complex Multiplication we refer to SHIMURA's book [78, Ch. 5]. As an old problem it was called "KRONECKER's Jugendtraum" and really it appears in HILBERT's programme as "Aufgabe" (KRONECKER's problem) preparing the 12-th problem itself. It has been essentially solved by TAKAGI [87] in 1921.

On the other hand SCHNEIDER, see [82], proved that j takes transcendental values at algebraic points on the upper half plane $\mathbb{H} = \{z \in \mathbb{C}; Im\, z > 0\}$ which are not singular. In analogy to the exponential function we can summarize the situation in the following manner. Let

$$\mathbb{H}_{sing} = \{\sigma \in \mathbb{H}; [\mathbb{Q}(\sigma) : \mathbb{Q}] = 2\}$$

be the set of *singular moduli.* Altogether one knows:

I. j has algebraic values on $\mathbb{H}_{sing}$;

I'. j takes transcendental values on $\mathbb{H}(\bar{\mathbb{Q}})\backslash\mathbb{H}_{sing}$, where $\mathbb{H}(\bar{\mathbb{Q}})$ denotes the set $\mathbb{H} \cap \bar{\mathbb{Q}}$ of algebraic numbers on the upper half plane;

II. the number-theoretic construction / quality / meaning of $j(\sigma)$, $\sigma \in \mathbb{H}_{sing}$.

In order to explain II. we need

Definition 4.1. Let M/L be a finite extension of number fields. M is called a *class field* of L, if M/L is an abelian extension. The HILBERT *class field* of L is the maximal unramified abelian field extension of L.

Theorem 4.2 (see [78]). *If $\sigma \in \mathbb{H}$ is a singular module, then $\mathbb{Q}(\sigma)(j(\sigma))$ is a class field of $\mathbb{Q}(\sigma)$. It is the* HILBERT *class field of $\mathbb{Q}(\sigma)$ if and only if the $\mathbb{Z}$-module $\mathbb{Z} + \mathbb{Z}\sigma$ is a (fractional) ideal of $\mathbb{Q}(\sigma)$, this means an $\mathcal{O}_{\mathbb{Q}(\sigma)}$-*module.

Remark 4.3. The HILBERT class field can be considered as a finite number theoretic analogon of universal coverings of curves, say. Let $\pi : \mathbb{H} \to C$ be the universal covering of the (smooth compact complex algebraic) curve C and $\mathcal{L}$ a line bundle on C. The pull back $\pi^*\mathcal{L}$ of $\mathcal{L}$ is a trivial line bundle on $\mathbb{H}$ (isomorphic to $\mathbb{H} \times \mathbb{C}$). Line bundles on C correspond to invertible $\mathcal{O}_C$*-sheaves.* We change over now from C to the "arithmetic curve" $\operatorname{Spec} \mathcal{O}_L$, L a number field. Invertible sheaves on $\operatorname{Spec} \mathcal{O}_L$ correspond to (fractional) ideals of L, and trivial ones to principal ideals. It is a classical origin of class field theory to look for a field extension H/L, which "trivializes" all L*-ideals* $\mathfrak{a}$ by inverse image "lifting": $\mathfrak{a} \mapsto \mathcal{O}_H\mathfrak{a}$, see [78]. The "Principal Ideal Theorem" of class field theory says that the HILBERT class field H of L has this property. For a modern proof and more details we refer to [57].

4.2 Complex Multiplication on Abelian Varieties

As we know one has two possibilities to introduce and work with abelian varieties A:

1) A is a torus $\mathbb{C}^g/\Lambda$, and there exists a RIEMANN *form* $E : \Lambda \times \Lambda \to \mathbb{Q}$. This is, by definition, a skew-symmetric non-degenerate $\mathbb{Z}$-bilinear form, also called *polarization.*
2) A is a compact complex algebraic variety with a commutative group structure. There exists a projective embedding $A \hookrightarrow \mathbb{P}^N$. Theta functions corresponding to the "lattice polarization" E are used to realize a projective embedding of $\mathbb{C}^g/\Lambda$. We refer to [55] for details. A *projective abelian variety* is an abelian variety together with a fixed projective embedding $A \hookrightarrow \mathbb{P}^N$. We also call them (projectively) *embedded abelian varieties.* Via images they can be identified with abelian subvarieties of projective spaces, that means with projective varieties with (abelian) group structure. Forgetting the group law for a moment we see that a projective abelian variety is defined by a system of homogeneous equations. We remark that projective embedding is an algebraic version of polarization (not the precise one, but sufficient for us at this moment).

Corresponding to the toroidal and the algebraic definition one discovers two aspects to consider "arithmetically defined abelian varieties". On the one hand there is the class of abelian varieties $\mathbb{C}^g/\Lambda$ with *arithmetic lattice* Λ, that means $\Lambda \subset L^g$, L a number field. On the other hand we have at our disposal on projective abelian varieties defined over a number field M, that means we can choose the defining equations with coefficients in M. The two arithmetic aspects are far away from being equivalent. For elliptic curves the problem of distinction leads immediately to the study of special values of the elliptic modular function j. In arbitrary dimension we look for abelian varieties which are arithmetically defined simultaneously in both senses. A condition implying arithmeticity in both senses would be helpful. This is our starting motivation for introducing abelian varieties with complex multiplication.

The theory of complex multiplication of abelian varieties in arbitrary dimension has been essentially founded by SHIMURA-TANIYAMA [79]. We follow mainly the monograph [48] of LANG, reproduce the basic results we need and give some proofs in order to prepare the reader to our applications of the theory.

Definition 4.4. *Let A be an abelian variety and End(A) its endomorphism ring.*

The $\mathbb{Q}$-algebra $\mathbb{Q}\otimes$ End(A) is called the *endomorphism algebra* or the *algebra of complex multiplication* of A.

We say that A has F*-multiplication,* if F is a number field and there exists an embedding $F \hookrightarrow \mathbb{Q}\otimes$ End(A) (as rings with 1). We call F a *multiplication field* of A.

A has *complex multiplication* (CM), if A has F-multiplication and $[F : \mathbb{Q}] = 2\dim A$. In this case A is called an (abelian) *CM-variety.*

Two abelian varieties A, B are called *isogeneous*, if there is a surjective morphism *(isogeny)* $A \to B$ with finite kernel. This relation is symmetric. We write $A \sim B$, if A and B are isogeneous.

The abelian variety A is called *simple*, if A doesn't contain any abelian subvariety of lower positive dimension. A necessary and sufficient criterion is that $\mathbb{Q} \otimes \mathrm{End}(A)$ is a division algebra, see [48]. For each abelian variety A there exists a decomposition

$$A \sim B_1 \times \ldots \times B_r ,$$

where the B_i's are simple abelian varieties. These decomposing factors are uniquely determined up to isogeny (POINCARÉ's Reducibility Theorem).

Proposition 4.5. *Let A be an abelian variety with F-multiplication. Then $[F : \mathbb{Q}] \leq 2 \cdot \dim A$. If A is a simple CM-variety and $[F : \mathbb{Q}] = 2 \cdot \dim A$, then F is uniquely determined up to isomorphy. It coincides with the endomorphism algebra $\mathbb{Q} \otimes$ End A in this case.*

Proof: Let F be a multiplication field of say $A, F = \mathbb{Q}(f)$. We can assume that $f \in \mathrm{End}(A)$. We have at our disposal faithful representations

$$\begin{aligned} R : \mathbb{Q} \otimes \mathrm{End}(A) &\longrightarrow \mathrm{Mat}_{g \times g}(\mathbb{C}) \\ R_\wedge : \mathrm{End}(A) &\longrightarrow \mathrm{End}_\mathbb{Z}(\wedge) \to \mathrm{Mat}_{2g \times 2g}(\mathbb{Z}). \end{aligned} \tag{4.1}$$

They come from commutative diagrams

$$\begin{array}{ccccccccc} 0 & \longrightarrow & \wedge & \longrightarrow & \mathbb{C}^g & \longrightarrow & A & \longrightarrow & 0 \\ & R_\wedge(f) \downarrow & & & \downarrow R(f) & & \downarrow f & & \\ 0 & \longrightarrow & \wedge & \longrightarrow & \mathbb{C}^g & \longrightarrow & A & \longrightarrow & 0 \end{array} \tag{4.2}$$

lifting elements $f \in \mathrm{End}(A)$. Let $\chi(T)$ be the characteristical polynomial of $R_\wedge(f)$. It is obvious that

$$\deg \chi(T) \leq 2g \quad \text{and} \quad \chi\,(R_\wedge(f)) = 0 ,$$

hence $[F : \mathbb{Q}] \leq 2g$.

Now assume that A is simple. We need two auxiliary results of algebra. The *centralizer* of $F \subseteq R$ in the ring R is denoted or defined by

$$Z_R(F) = \{r \in R;\ rf = fr \text{ for all } f \in F\} .$$

4.6. Let R be a division algebra with center Z and F a subfield of R such that $Z \subseteq F \subseteq R$ and $Z_R(F) = F$. Then $[F : Z] = d$, where $d = [R : Z]$.

4.7. Let F be a simple k-algebra, $h = [F : k]$. Assume that there is a faithful representation of F in the k-vector space V of dimension h. Then $Z_{\mathrm{End}_k V}(F) = F$.

Remark 4.8. For the proof of 4.6 and 4.7 we refer to REINER's monograph [63]. It is easy to derive 4.6 and 4.7 from the following fundamental result:

Theorem 4.9 ([63], (Theorem (7.11), Corollary (7.13)). *Let $K \subseteq B \subseteq A$, where B is a simple subring of the central simple K-*algebra *A, A finite over K. Then it holds that*

$$[B:K]\cdot[Z_A(B):K]=[A:K].$$

Now we apply 4.7 to the multiplication field F of $A, k=\mathbb{Q}$ and the faithful representation $R_\wedge$ in $V=\mathbb{Q}^h$, $h=2g$, see (4.1). It follows immediately that F coincides with its centralizer in the division algebra $R=\operatorname{End}A$. Applying 4.6 we get $[F:Z]=d$. We show that $d=1$. We have

$$2g=[F:\mathbb{Q}]=[F:Z]\cdot[Z:\mathbb{Q}]=d[Z:\mathbb{Q}]\,. \tag{4.3}$$

Since $\mathbb{Q}^{2g}$ is a module over the division algebra $R=\operatorname{End}A$ and $[R:\mathbb{Q}]=d^2[Z:\mathbb{Q}]$ it follows that $d^2[Z:\mathbb{Q}]$ divides $2g$. This is only compatible with (4.3) if $d=1$. But this means $Z=F$. Consequently

$$\operatorname{End}A=Z_R(Z)=Z_R(F)=F\,,$$

where the latter identity comes from 4.7 again. The proposition is proved. □

4.3 Types of Complex Multiplication

Let A be an abelian CM-variety of dimension g with multiplication field F, $[F:\mathbb{Q}]=2g$. We have an exact sequence

$$0\longrightarrow\wedge\longrightarrow T_A\longrightarrow A\longrightarrow 0\,, \tag{4.4}$$

where $T_A\cong\mathbb{C}^g$ denotes the tangent space of A (at 0). The representation R of F in T_A, see (4.1), splits into g one-dimensional representations because F is commutative. We identify them with characters $\varphi_i:F\to\mathbb{C}$, $i=1,\dots,g$. These are embeddings into $\mathbb{C}$. So our representation R of F in T_A is equivalent to the more explicit one

$$\begin{aligned} D_\phi : F &\longrightarrow \operatorname{Mat}_{g\times g}(\mathbb{C}) \\ f &\longmapsto D_\phi(f)=\operatorname{diag}\left(\varphi_1(f),\dots,\varphi_g(f)\right)\,, \end{aligned} \tag{4.5}$$

where ϕ is the sum of representations $\sum_{i=1}^{g}\varphi_i$, also understood as sum of embeddings

$$\phi=\sum_{i=1}^{g}\varphi_i : F\hookrightarrow\mathbb{C}^g\,,\quad f\mapsto {}^t\left(\varphi_1(f),\dots,\varphi_g(f)\right)\,. \tag{4.6}$$

Conversely, let L be a *lattice in* F, that means a $\mathbb{Z}$-submodule of rank $[F:\mathbb{Q}]$ of F, $[F:\mathbb{Q}]=2g$ and ϕ as in (4.6) composed by g field embeddings $\varphi_i : F \hookrightarrow \mathbb{C}$. We consider the torus $\mathbb{C}^g/\phi(L)$ assuming that the $\mathbb{Z}$-rank of $\phi(L)$ is not smaller than that of L. Using D_ϕ defined in (4.5) we see that $\mathbb{C}^g/\phi(L)$ has F-multiplication in a canonical manner. We already introduced ϕ as an invariant of an abelian CM-variety with F-multiplication via tangent spaces. This is also possible for tori. Therefore ϕ is an invariant of $\mathbb{C}^g/\phi(L)$ and its F-multiplication.

Definition 4.10. An abelian CM-variety A is called of *model type* (L,ϕ), if A is isomorphic to $\mathbb{C}^g/\phi(L)$ for a suitable lattice $L \subset F$ and ϕ as above.

Let $\varphi : F \hookrightarrow \mathbb{C}$ be a field embedding. Then we denote its complex conjugate by $\bar{\varphi}$. It is defined by $\bar{\varphi}(f) = \overline{\varphi(f)}$.

Lemma 4.11. *If A is an abelian CM-variety of type (L,ϕ) as above, then the summands $\varphi_i : F \hookrightarrow \mathbb{C}$ of ϕ satisfy the following conditions:*

$$\varphi_i \neq \varphi_j \text{ for } i \neq j\,, \quad \varphi_i \neq \bar{\varphi}_j\,, \quad i,j = 1,\ldots,g\,. \tag{4.7}$$

Especially, the multiplication field F is a totally imaginary number field.

Proof: By assumption there is an isomorphism $A \cong \mathbb{C}^g/\phi(L)$ and A has a period matrix satisfying the RIEMANN period relations. This property transfers to the matrix

$$\Pi = \begin{pmatrix} \varphi_1(\ell_1) & \ldots & \varphi_1(\ell_{2g}) \\ \cdot & \ldots & \cdot \\ \varphi_g(\ell_1) & \ldots & \varphi_g(\ell_{2g}) \end{pmatrix} \tag{4.8}$$

where $\ell_1,\ldots,\ell_{2g}$ is a $\mathbb{Z}$-basis of L such that the columns generate the lattice $\phi(L)$. It suffices to show that the $2g \times 2g$-matrix $\binom{\Pi}{\bar{\Pi}}$ is regular. But this follows from the transferred period relations, namely

$$\begin{pmatrix} \Pi \\ \bar{\Pi} \end{pmatrix} Q({}^t\Pi \mid {}^t\bar{\Pi}) = \left(\begin{array}{c|c} \Pi Q^t\Pi & \Pi Q^t\bar{\Pi} \\ \hline \bar{\Pi} Q^t\Pi & \bar{\Pi} Q^t\bar{\Pi} \end{array}\right) = \left(\begin{array}{c|c} 0 & H \\ \hline \bar{H} & 0 \end{array}\right)$$

with a suitable skew-symmetric regular $Q \in Gl_{2g}(\mathbb{Q})$ and regular H. The lemma is proved. □

Definition 4.12. With the notations of 4.6 we call the pair (F,ϕ) a *CM-type*, if the summands φ_i of ϕ satisfy the condition (4.7). Especially F has to be a totally imaginary number field.

It is interesting to ask for all CM-types coming from abelian varieties. We already verified in section 2 that each abelian CM-variety (A,ι), $\iota : F \hookrightarrow \mathbb{Q}\otimes \operatorname{End} A$ defines a pair (F,ϕ) in a unique manner. We want to check that this pair is really a CM-type in the sense of the above definition. This is part of the next

Proposition 4.13. *Let A be an abelian variety of dimension g. If A is of model type (L, ϕ), then it has complex multiplication of CM-type (F, ϕ), where $F = \mathbb{Q} \otimes L$. Conversely, if (A, ι) is an abelian CM-variety with complex multiplication $\iota : F \hookrightarrow \mathbb{Q} \otimes \operatorname{End} A$, then A is of model type (L, ϕ) for a suitable $\mathbb{Z}$-lattice L of F. Especially the pair (F, ϕ) is a CM-type.*

Proof: The first assertion comes from Lemma 4.11, also the last one, if we have proved the second statement.
So we assume that A has F-multiplication and $[F : \mathbb{Q}] = 2g$ with associated pair (F, ϕ), $\phi = \sum_{i=1}^{g} \varphi_i$. The embeddings φ_i are the weights of the representation R of F in the tangent space T_A. We choose eigenvectors $\mathfrak{a}_i$ in T_A corresponding to φ_i. With obvious notations it holds that

$$R(f)\mathfrak{a}_i = \varphi_i(f)\mathfrak{a}_i \quad \text{for } f \in F, \quad i = 1, \ldots, g.$$

The representation R is equivalent to the representation D_ϕ defined in (4.5). The equivalence is realized by the commutative diagrams

$$\begin{array}{ccc} & R(f) & \\ T_A & \longrightarrow & T_A \\ \kappa \updownarrow & & \downarrow \kappa \\ \mathbb{C}^g & \longrightarrow & \mathbb{C}^g \\ & D_\phi(f) & \end{array}$$

where κ is the linear coordinate map sending the $\mathfrak{a}_i$'s to the members $\mathfrak{n}_i$ of the canonical basis of $\mathbb{C}^g$, respectively. So we have

$$(*) \qquad \kappa(f\mathfrak{a}) = D_\phi(f)\kappa(\mathfrak{a}) \quad \text{for } f \in F, \ \mathfrak{a} \in T_A.$$

Let Λ be the kernel lattice of the canonical uniformization $T_A \longrightarrow A$ of A, see (4.4). For $\lambda \in \Lambda \backslash 0$ we define

$$L_\lambda = [\Lambda : \Lambda]_F = \{f \in F; f\lambda \in \Lambda\}.$$

We claim that

$$(**) \qquad L_\lambda \cdot \lambda = \Lambda.$$

First we remark that

$$\text{(i)} \qquad \mathbb{Q} \cdot L_\lambda = F.$$

Namely, for $f \in F \subseteq \mathbb{Q} \otimes \operatorname{End} A$ we find $m \in \mathbb{Z} \backslash 0$ such that $mf \in \operatorname{End} A$. The elements of $\operatorname{End} A$ act on Λ, hence $mf \cdot \lambda \in \Lambda$, $mf \in L_\lambda$ and finally $f \in \mathbb{Q} \cdot L_\lambda$. Now we check

$$\text{(ii)} \qquad F \cdot \lambda = \mathbb{Q}\Lambda.$$

The $\mathbb{Q}$-linear map $F \longrightarrow \mathbb{Q}\Lambda$, $f \mapsto f\lambda$, is injective because the non-trivial elements of F are invertible and $1 \cdot \lambda = \mathrm{id} \circ \lambda = \lambda \neq 0$. Both, F and $\mathbb{Q}\Lambda$, have $\mathbb{Q}$-dimension $2g$. This proves (ii).

Now we are able to verify $(**)$. The inclusion $\subseteq$ comes from the definition of L_λ. Let now x be an element of Λ. By (ii) we can identify $F \cdot \lambda$ and $\mathbb{Q}\Lambda$ with $F \subseteq \mathbb{Q} \otimes \mathrm{End}\, A$. We set $y = x/\lambda$ in this sense and find that $y \in L_\lambda$. Thus x can be written as $y \cdot \lambda$ as we need it for the proof of $(**)$.
We finish the proof of the proposition by showing that $A \cong \mathbb{C}/\phi(L_\lambda)$ for a suitable $\lambda \in \Lambda$. For this purpose we construct a commutative diagram

$$\begin{array}{ccccccccc}
0 & \longrightarrow & \Lambda & \longrightarrow & T_A & \longrightarrow & A & \longrightarrow & 0 \\
 & & \kappa \downarrow & & \downarrow \kappa & & \uparrow & & \\
 & & \kappa(\Lambda) & \longrightarrow & \mathbb{C}^g & & | & & \\
 & & \uparrow & & \uparrow S & & | & & \\
0 & \longrightarrow & \phi(L_\lambda) & \longrightarrow & \mathbb{C}^g & \longrightarrow & \mathbb{C}^g/\phi(L_\lambda) & \longrightarrow & 0
\end{array}$$

with vertical isomorphisms. The left upper part is clear by the definition of κ. In order to construct S we apply $(*)$ to $\mathfrak{a} = \lambda$ setting $\kappa(\lambda) = {}^t(\wr'_1, \dots, \wr'_g)$:

$$(***) \qquad \kappa(f\lambda) = D_\phi(f)\kappa(\lambda) = \mathrm{diag}\,(\wr'_1, \dots, \wr'_g)^t(\varphi_1(f), \dots, \varphi_g(f)) = S \cdot \phi(f).$$

For suitable $\lambda \in \Lambda$ we can be sure that all $\wr'_i \neq 0$. Then the diagonal matrix S is regular and defines, with the same notation, the isomorphism in diagram (4.9) we look for. The S-image of $\phi(L_\lambda)$ is $\kappa(\Lambda)$. This follows immediately from $(***)$ and $(**)$:

$$S\phi(L_\lambda) = \kappa(L_\lambda) = \kappa(\Lambda).$$

The vertical isomorphisms of the left-hand side in (4.9) define a model isomorphism for A on the right. The proposition is proved. □

Corollary 4.14. *Each abelian CM-variety A of dimension g is isomorphic to a torus $\mathbb{C}^g/\Lambda$ with an arithmetic lattice Λ.*

Corollary 4.15. *There are only countable many isomorphy classes of abelian CM-varieties of dimension g.*

Taking into account additionally principal polarizations we prove more:

Proposition 4.16 (Rigidity). *There are only countable many isomorphism classes of principally polarized abelian CM-varieties of dimension g.*

Proof: By 4.14 we can assume that $A = \mathbb{C}^g/\Lambda$, $\Lambda \subset \overline{\mathbb{Q}}^g$. A principal polarization $E : \Lambda \times \Lambda \longrightarrow \mathbb{Z}$ is completely determined by the knowledge of a symplectic basis $\lambda_1, \dots, \lambda_{2g}$ of Λ. The corresponding "period matrix" Π with columns $\lambda_i, i = 1, \dots, 2g$, belongs to $\mathrm{Mat}_{g \times 2g}(\overline{\mathbb{Q}})$. This is a countable set because $\overline{\mathbb{Q}}$ is. □

Remark 4.17. Rigidity means that (principally) polarized abelian CM-varieties "have no moduli". In other words there are only trivial algebraic families $\mathcal{A} \longrightarrow S$ of abelian CM-varieties.

On the other hand it is not difficult to see that moduli points of principally polarized abelian CM-varieties of dimension form a dense (countable) set in the moduli space $\mathcal{A}_g(\mathbb{C})$. This is also true for any other (fixed) type of polarization.

We would like to know how much isomorphism classes of abelian varieties of fixed type (F, ϕ) exist. We know that for each CM-type there exist tori with complex multiplication of the given type by our model constructions. There exists an elementary number theoretic criterion deciding whether a torus of type (F, ϕ) is an abelian variety or not. This depends only on the type, see Proposition 4.27 below. So there will be no confusion if we restrict ourselves to CM-types (F, ϕ) having an (only) abelian variety model(s). As first step we restrict the model types (L, ϕ) to a smaller representing class.

Definition 4.18. With the notations of (4.6), 4.10 we call (L, ϕ) an *ordered model type*, if the following conditions are satisfied:

(i) L is an $\mathcal{O}$-module for $\mathcal{O} = [L : L]_F = \{f \in F; fL \subseteq L\}$,
(ii) $\mathcal{O} = F_\cap \operatorname{End} A = \iota^{-1}(\operatorname{End} A)$,

where $\iota : F \longrightarrow \mathbb{Q} \otimes \operatorname{End} A$ is the CM-embedding.

From the definition it is clear that $\mathcal{O}$ is an order of F, that means a $\mathbb{Z}$-lattice of F with induced ring structure. For a better symbolic distinction we write $\mathfrak{a}$ instead of L if the conditions (i), (ii) are satisfied. If $\mathcal{O}$ coincides with the ring $\mathcal{O}_F$ of integers of F, then $\mathfrak{a}$ is nothing else than a fractional ideal of F.

Lemma 4.19. *Each abelian CM-variety A of dimension g is of ordered model type $(\mathfrak{a}, \phi)$ for suitable $\mathfrak{a} \subset F$.*

Proof: Assume that A has complex multiplication of type (F, ϕ). We consider F as subspace of $\mathbb{Q} \otimes \operatorname{End} A$ and set $\mathcal{O} = F_\cap \operatorname{End} A$. Obviously $\mathcal{O}$ is an order in F. In the proof of Proposition 4.13 we have verified that A is isomorphic to $\mathbb{C}^g/\phi(L_\lambda)$ with the notations of diagram (4.9). We check that L_λ is an $\mathcal{O}$-module: For $\omega \in \mathcal{O}, 1 \in L_\lambda$ one finds that

$$(\omega l)\lambda = \omega(l\lambda) \in \omega(L_\lambda \cdot \lambda) = (\text{by } (**))\, \omega\Lambda \subseteq \Lambda.$$

hence $\omega l \in L_\lambda$. Setting $\mathfrak{a} = L_\lambda$ it remains to verify condition (i), that means $[\mathfrak{a} : \mathfrak{a}]_F = \mathcal{O}$. Since $\mathfrak{a}$ is an $\mathcal{O}$-module we have only to check the inclusion $\subseteq$. Let $f \in [\mathfrak{a} : \mathfrak{a}]_F$. Then $fl_\lambda \subseteq L_\lambda$, hence $fL_\lambda \cdot \lambda \subseteq L_\lambda \cdot \lambda$. This means $f\Lambda \leq \Lambda$ by $(**)$ again. So f defines a commutative diagram (4.2). Consequently f belongs to $(\operatorname{End} A)_\cap F$, which is $\mathcal{O}$ by definition.

Conversely, if $(\mathfrak{a}, \phi)$ is a model type, $\mathcal{O}$ an order in F and $\mathfrak{a}$ an $\mathcal{O}$-module satisfying $[\mathfrak{a} : \mathfrak{a}]_F = \mathcal{O}$, then the standard model $A = \mathbb{C}^g/\phi(\mathfrak{a})$ is of ordered model type $(\mathfrak{a}, \phi)$. Using the standard multiplication introduced in (4.5) we have to show that the above condition (ii) is satisfied.

For $\omega \in \mathcal{O}$ we have $\omega\mathfrak{a} \subseteq \mathfrak{a}$, hence

$$\omega \circ \phi(\mathfrak{a}) = D_\phi(\omega)\phi(\mathfrak{a}) = {}^t(\varphi_1(\omega) \cdot \varphi_1(\mathfrak{a}), \dots, \varphi_g(\omega) \cdot \varphi_g(\mathfrak{a})) = \phi(\omega\mathfrak{a}) \in \phi(\mathfrak{a}),$$

hence $\mathcal{O} \subseteq F_\cap \mathrm{End}\, A$. For the inverse inclusion we take $\omega \in F_\cap \mathrm{End}\, A$. Successively we get

$$\omega \circ \phi(\mathfrak{a}) \subseteq \phi(\mathfrak{a}), \quad \phi(\omega\mathfrak{a}) \subseteq \phi(\mathfrak{a}), \quad \varphi_i(\omega\mathfrak{a}) \subseteq \varphi_i(\mathfrak{a}), \quad \omega\mathfrak{a} \subseteq \mathfrak{a}$$

and finally $\omega \in \mathcal{O}$.

Now we are going to list isomorphy classes of abelian CM-varieties of fixed type (F, ϕ). For any number field F and any order $\mathcal{O}$ of F we consider the class of $\mathcal{O}$-modules $\mathfrak{a} \subseteq F$ with the property $[\mathfrak{a} : \mathfrak{a}]_F = \mathcal{O}$. The set of $\mathcal{O}$-isomorphy classes of them is denoted by $cl(\mathcal{O})$ and its cardinality by $h(\mathcal{O})$. For $\mathcal{O} = \mathcal{O}_F$ this is nothing else than the class group or the class number, respectively, of F.

The set $cl_p(\mathcal{O})$ is defined by the finer principal equivalence relation: $\mathfrak{a} \sim \mathfrak{b}$ if there exists $f \in F$ such that $\mathfrak{b} = f\mathfrak{a}$. Obviously it holds that $\sharp cl(\mathcal{O}) \leq \sharp cl_p(\mathcal{O})$. The latter number, hence also $cl(\mathcal{O})$, is finite. A proof can be found in BOREVIČ-SHAFAREVIČ's textbook [11], II.6, Theorem 3. We get the following Finiteness Theorem in this way:

Proposition 4.20. *Let (F, ϕ) be a complex multiplication type of an abelian variety, $\mathcal{O}$ an order in F. The number $h_\phi(\mathcal{O})$ of isomorphy classes of abelian CM-varieties of type (F, ϕ) with $\mathcal{O} \cong F_\cap \mathrm{End}\, A$ is finite. More precisely it holds that*

$$h(\mathcal{O}) = \sharp cl(\mathcal{O}) \leq h_\phi(\mathcal{O}) \leq h_p(\mathcal{O}) = \sharp cl_p(\mathcal{O}).$$

Proof: We have only to compare standard models $\mathbb{C}^g/\phi(\mathfrak{a})$, where $\mathfrak{a}$ runs through the representants of $cl(\mathcal{O})$. If $\mathfrak{b} = n\mathfrak{a}$ for a natural number n then the isomorphy of $\mathbb{C}^g/\phi(\mathfrak{a})$ and $\mathbb{C}^g/\phi(\mathfrak{b})$ is obvious. So we can restrict us in the more general situation $\mathfrak{b} = f\mathfrak{a}$ to the case $f = \omega \in \mathcal{O} \subseteq \mathrm{End}\, A$. Then $R(\omega)$ induces an isomorphism $\mathbb{C}^g/\phi(\mathfrak{a}) \cong \mathbb{C}^g/\phi(\mathfrak{b})$. This proves the second inequality, hence the finiteness result.

If $\alpha : A \longrightarrow B$ is an isomorphism of two abelian varieties in our class, then it lifts to $R(\alpha) : T_A \xrightarrow{\sim} T_B$. The endomorphism rings End A and End B are identified along α by means of commutative diagrams

$$\begin{array}{ccc} A & \xrightarrow[\sim]{\alpha} & B \\ \omega \downarrow & & \downarrow \omega' \\ A & \xrightarrow[\alpha]{\sim} & B \end{array}$$

The diagram lifts commutatively to the tangent level with morphisms $R(\alpha), R(\omega)$. Restricting to kernel lattices Λ_A, Λ_B and $\omega \in \mathcal{O}$ we see that $\mathcal{O}$-multiplication is compatible with the lattice isomorphism $R(\alpha) : \Lambda_A \xrightarrow{\sim} \Lambda_B$. Consequently, the first inequality is also verified. □

4.4 Transformation of Constants

We remember that from our point of view a (complex) abelian variety is a complex torus, which can be embedded into a (complex) projective space $\mathbb{P}^N(\mathbb{C})$. A projective abelian variety is an abelian variety together with a fixed such embedding. Forgetting the multiplication, which is unique up to translation of the neutral point 0, it can be described (in many ways) by a system of homogeneous equations

$$F_1 = \ldots = F_r = 0, \quad F_i \text{ (homogeneous) } \in \mathbb{C}[X_0, X_1 \ldots, X_N], \tag{4.10}$$

generating the corresponding homogeneous prime ideal.

Now let $\mu \in \operatorname{Aut}\mathbb{C}$ be a field isomorphism of the complex numbers and V an arbitrary projective variety defined by (4.10), say. Then we define V^μ by

$$V^\mu : F_1^\mu = \ldots = F_r^\mu = 0,$$

where

$$F^\mu = \sum \mu(a_{i_0 \ldots i_N}) X_0^{i_0} \cdots X_N^{i_N} \quad \text{for} \quad F = \sum a_{i_0 \ldots i_N} X_0^{i_0} \cdots X_N^{i_n}.$$

In general V and V^μ are not isomorphic (as complex algebraic varieties). So, applying $\operatorname{Aut}\mathbb{C}$, we obtain in this way from V a lot of new varieties by *transformations of constants.*

The application of $\operatorname{Aut}\mathbb{C}$ is functorial in the following sense: We correspond to any algebraic (that means rational) morphism $\alpha : V \longrightarrow W$ the morphism $\alpha^\mu : V^\mu \longrightarrow W^\mu$ by transformation of constants again applied to the rational functions defining α. For two morphisms α and $\beta : W \longrightarrow Z$ it holds that $(\beta \circ \alpha)^\mu = \beta^\mu \circ \alpha^\mu$.

Especially, if α is an isomorphism, then the transforms V^μ and W^μ are isomorphic again. Consequently,

4.21. $\operatorname{Aut}\mathbb{C}$ acts on the isomorphy classes of projective varieties.

In this sense the transform $A^\mu, \mu \in \operatorname{Aut}\mathbb{C}$, of an abelian variety A is a well-defined abelian variety not depending on the special choice of projective embedding, up to isomorphy.

Remark 4.22. At this point the scheme language is useful even in the case of complex projective varieties. Abstract complex varieties in the sense of schemes appear as morphisms $V \longrightarrow c$ patching together affine parts $\operatorname{Spec} R \longrightarrow c = \operatorname{Spec}\mathbb{C}$ corresponding to ring embeddings $\mathbb{C} \longrightarrow R$. We write $V \longrightarrow \mathbb{C}$ or $V/\mathbb{C}$ instead of the more precise use of $c = \operatorname{Spec}\mathbb{C}$. Scheme theory makes free from the use of special projective embeddings. The action of $\operatorname{Aut}\mathbb{C}$ on the class of schemes over $\mathbb{C}$ can be introduced by means of fibre products of scheme theory. For $\mu \in \operatorname{Aut}\mathbb{C}$ and $V/\mathbb{C}$ as above the transform V^μ is uniquely determined up to isomorphy by the Cartesian (fibre product) diagram

$$\begin{array}{ccc} V^\mu & \longrightarrow & V \\ \downarrow & & \downarrow \\ \mathbb{C} & \underset{\mu}{\xrightarrow{\sim}} & \mathbb{C} \end{array} \tag{4.11}$$

A $\mathbb{C}$-morphism $W/\mathbb{C} \longrightarrow V/\mathbb{C}$ is a commutative diagram (4.11) with $W, \mathrm{id}_{\mathbb{C}}$ instead of V^μ or μ, respectively. It is easy to verify functoriality of the application of Aut $\mathbb{C}$ in the above sense.

Now let k be a subfield of $\mathbb{C}$, A an abelian variety with *projective* model V (the image of a projective embedding of A). If (4.10) is a system of defining equations for V and all coefficients of all polynomials F_i belong to k, then we call V a *projective k-model* of A. In the more general scheme language a k-model of A is a scheme V_k/k such that $V_{\mathbb{C}} = \operatorname{Spec}\mathbb{C} \times_k V_k$ is $\mathbb{C}$-isomorphic to $A/\mathbb{C}$.

Lemma 4.23. *If a projective variety V has a $\overline{\mathbb{Q}}$-model, then the set of isomorphy classes of transformed varieties*

$$\{V^\mu; \mu \in \operatorname{Aut}\mathbb{C}\}/\,\mathbb{C} - \text{isom.} \tag{4.12}$$

is finite.

Proof: We can assume that V is defined by the equations (4.10) and the polynomials F_i have coefficients in $\overline{\mathbb{Q}}$. Then we have to apply $\mu \in \operatorname{Aut}\mathbb{C}$ to the coefficients to get V^μ. But for any algebraic number α the orbit $(\operatorname{Aut}\mathbb{C})\alpha$ is finite. □

Theorem 4.24 (SHIMURA-TANIYAMA) [79]; see [48], Proposition V.1.1). *Each abelian CM-variety has a projective $\overline{\mathbb{Q}}$-model.*

We give only a proof for the Jacobians of PICARD curves. If A is an abelian CM-variety, then also A^μ is for any $\mu \in \operatorname{Aut}\mathbb{C}$ because F-multiplication transfers to F^μ-multiplication on A^μ by functoriality. From 4.15 it follows that

$$\{A^\mu; \mu \in \operatorname{Aut}\mathbb{C}\}/\text{iso} \qquad \text{is countable.} \tag{4.13}$$

Here A can be understood also as projective (polarized) abelian variety by Proposition 4.16.

Now we consider $A = J(C)$, C a smooth PICARD curve, such that $J(C)$ has complex multiplication. Shortly we call them *CM*-PICARD *curves.* Canonical principal polarizations define projectively embedded abelian threefolds Jac C.

4.25. For a CM-PICARD curve $C : Y^3 = p_4(X)$ the set of isomorphy classes $\{C^\mu; \mu \in \operatorname{Aut}\mathbb{C}\}/\text{iso}$ is countable.

Namely, by functoriality it holds that $\operatorname{Jac}(C)^\mu = \operatorname{Jac}(C^\mu)$. The TORELLI theorem embeds the set of isomorphy classes 4.25 into that of 4.13 with $A = Jac\,(C)$. This proves 4.25. □

Now write $p_4(X) = \prod_{i=1}^{4} (X - \xi_i)$, $\sum \xi_i = 0$. Then it follows that the set of moduli points $\{(\xi_1^\mu : \xi_2^\mu : \xi_3^\mu : \xi_4^\mu); \mu \in \operatorname{Aut}\mathbb{C}\}$ is countable, hence (say $\xi_1 \neq 0$) $\{((\xi_2/\xi_1)^\mu, (\xi_3/\xi_4)^\mu, (\xi_4/\xi_1)^\mu); \mu \in \operatorname{Aut}\mathbb{C}\}$ is countable. But this can only happen, if ξ_i/ξ_1 is not a transcendental number for $i = 2, 3, 4$. This means that C has a $\overline{\mathbb{Q}}$-model. But then the algebraic construction of Jacobian varieties of curves D of genus g (birational equivalent to D^g/S_g) shows that also Jac(C) has a $\overline{\mathbb{Q}}$-model. □

4.5 SHIMURA Class Fields

We outline further basic tools we need from the SHIMURA-TANIYAMA theory of complex multiplication of (complex) abelian varieties in order to construct class fields by means of singular values of the PICARD modular theta functions following the book [48] of S. LANG. Next we have to introduce the reflex fields. Fixing notations we let F be a totally imaginary number field of absolute degree $2g$ and ϕ a choice of g embeddings $\varphi_i : F \longrightarrow \mathbb{C}$ pairwise different and not conjugated to each other. We write $\phi = \phi_F = \sum_{i=1}^{g} \varphi_i$ as above.

If M/F is a finite field extension, then we can lift ϕ_F to

$$\phi_M = \sum_{i=1}^{g} \sum \{\text{all extensions of } \varphi_i \text{ to } M\} .$$

So we get a *lifting of CM-types* $(F, \phi_F) \mapsto (M, \phi_M)$. We set

$$\mathrm{Stab}\,(\phi) = \{\mu \in \mathrm{Aut}\,\mathbb{C}; \mu \circ \phi = \phi\} . \tag{4.14}$$

Now assume that M/F as above is a GALOIS extension. Then the *reflex field* F' of (F, ϕ) is defined as fixed field

$$F' = M^{\mathrm{Stab}\,(\phi)} = \mathbb{Q}(Tr_\phi(F)), \tag{4.15}$$

where $Tr_\phi : F \longrightarrow F'$ denotes the *type trace* defined by $Tr_\phi(f) = \sum_{i=1}^{g} \varphi_i(f)$. For the second identity in (4.15) we refer to [48], I.5. With $\psi' = \psi^{-1}, \psi$ understood as automorphism of M, we set $\phi'_M = \sum_{\psi \in \phi_M} \psi'$. One can show that the type (M, ϕ'_M) is the lift of a uniquely determined primitive type (F', ϕ') which is called the *reflex type* of (F, ϕ). A type is called *primitive*, if it is not lifted from a lower field. If the starting type (F, ϕ) is primitive, then the reflex (F'', ϕ'') of its reflex (F', ϕ') coincides with (F, ϕ). In general, the double reflex field F'' is contained in F. Altogether we describe the situation in the following diagram

$$\begin{array}{ccccccc} & M & & & \phi_M & \mapsto & \phi'_M = \sum \psi'_j \\ \mathrm{Gal} \diagup & & \diagdown & & \uparrow & & \downarrow \\ F & & F', & \phi = \phi_F & & & \phi'_{F'} = \phi' \end{array} \tag{4.16}$$

On this place we are able to explain which CM-types (F, ϕ) can occur as types of abelian CM-varieties. Each abelian variety A can be decomposed up to isogeny into a product of simple abelian varieties. Simple abelian varieties are the indecomposable ones in this sense. If A is an abelian CM-variety, then the isogeny decomposition of A into simple abelian varieties is a power $A \approx B \times \ldots \times B$. The multiplication type (F, ϕ) of A is lifted from a uniquely determined multiplication type (E, ψ) of B. Especially, the simple factor B of A is a CM-variety. The corresponding CM-algebra (Proposition 4.5) $\mathbb{Q} \otimes \operatorname{End} B$ is isomorphic to E. Looking back to A one checks easily that

4.26. *The CM-algebra $\mathbb{Q} \otimes \operatorname{End} A$ of an abelian CM-variety $A \approx B \times \ldots \times B, B$ simple, is isomorphic to the matrix algebra $\operatorname{Mat}_s(E)$, where s denotes the number of the decomposing factors B of A and E is the CM-field $\mathbb{Q} \otimes \operatorname{End} B$.*

For the proof of the property of *abelian CM-types* to be lifted from primitive types we refer to [48], Theorem I.3.4. The inverse statement is also true by [48], Theorem I.4.4. We change over to tori in order to have the following criterion:

Proposition 4.27. *The standard torus $\mathbb{C}^g/\phi(\mathfrak{a})$ of given multiplication type (F, ϕ) is an abelian variety iff this type is lifted from a primitive type (E, ψ).*

Fixing F and ϕ, the *type norm* N_ϕ or the *reflex norm* $N' = N_{\phi'}$ are respectively defined by

$$N_\phi : F \longrightarrow F', \quad f \mapsto \prod_{i=1}^{g} \varphi_i(f),$$

$$N_{\phi'} = N' : F' \longrightarrow F'' \hookrightarrow F \quad \text{analogeously.}$$

Both, N_ϕ and N', can be extended to the idele groups of the fields F or F', respectively:

$$N_\phi : \mathbb{A}_F^* \longrightarrow \mathbb{A}_{F'}^*, \quad N' : \mathbb{A}_{F'}^* \longrightarrow \mathbb{A}_F^*.$$

Now we are well-prepared to define the SHIMURA class fields mentioned above. For this purpose we let $\mathfrak{a}$ be a $\mathbb{Z}$-lattice in F. The absolute norm of ideles s is denoted by $\mathbb{N}(s)$. Now we define the idele group $U(\phi, \mathfrak{a}) \subseteq \mathbb{A}_{F'}^*$, of the *extended type* $(F, \phi, \mathfrak{a})$ by

$$U(\phi, \mathfrak{a}) = \{s \in \mathbb{A}_{F'}^*;\ N'(s^{-1})\mathfrak{a} = \beta\mathfrak{a},\ \beta\bar{\beta} = \mathbb{N}(s^{-1}) \in \mathbb{Q} \text{ for a suitable } \beta \in F\}\,. \tag{4.17}$$

We remark that the multiplication of an idele $t \in \mathbb{A}_F^*$ with $\mathfrak{a}$ is defined componentwise on the finite part $t_{\text{fin}} = (t_p) \in \mathbb{A}_{F,\text{fin}}^*$: There is a unique $\mathbb{Z}$-lattice $t\mathfrak{a}$ in F with local components

$$(t\mathfrak{a})_p = t_p\mathfrak{a}_p \text{ for all } p \in \operatorname{Spec} \mathbb{Z}, \ (\text{where } \mathfrak{a}_p = \mathbb{Z}_p \otimes \mathfrak{a}).$$

Now we apply global abelian class field theory in order to define $Sh(\phi, \mathfrak{a})$ as class field of the reflex field F'. For details we refer to the monograph [57] of NEUKIRCH.

Let M be a finite abelian field extension of F'. Then there is an exact sequence

$$1 \longrightarrow U/F'^{\times} \longrightarrow \mathbb{A}_{F'}^*/F'^{\times} \overset{(.,M/F')}{\longrightarrow} \operatorname{Gal}(M/F') \longrightarrow 1 \tag{4.18}$$

where $(t, M/F'), t \in \mathbb{A}_{F'}^*$ is the global norm rest symbol locally defined by FROBENIUS automorphisms. The idele group $U = U_M$ is equal to the extended norm group $N_{M/F'}(\mathbb{A}_M^*)F'$. Conversely, if U is a cofinite subgroup of the idele group $\mathbb{A}_{F'}^*$ containing F', then there exists a unique finite abelian extension $M_U = M$, the *class field of* F' *belonging to* U, such that the above sequence (4.18) is exact. So there is a biunivoque correspondence (reciprocity):

$$U_M = U \longleftrightarrow M = M_U.$$

Now we take the projective limit of our finite abelian groups $\mathrm{Gal}(M/F')$ along all finite abelian extensions M of F'. In this way we obtain the GALOIS group $\mathrm{Gal}(F'^{ab}/F')$ of the maximal abelian extensions F'^{ab} of F'. The norm rest maps in (4.18) yield an injective map $(.,F') : \mathbb{A}_{F'}^*/F'^{\times} \longrightarrow \mathrm{Gal}(F'^{ab}/F')$. Via the norm rest symbols (s,F') the idele group $\mathbb{A}_{F'}^*$ acts on F'^{ab}, and the finite abelian extension fields M appear as fixed fields of the corresponding subgroup U of $\mathbb{A}_{F'}^*$. So we can write

$$M_U = (F'^{ab})^{(U,F')} = \mathbb{C}^{\widetilde{(U,F')}}, \tag{4.19}$$

where $\widetilde{(U,F')}$ denotes the group of all extensions of elements $(s,F') \in \mathrm{Gal}(F'^{ab}/F')$, $s \in U$, to automorphisms of $\mathbb{C}$.

Definition 4.28. The SHIMURA *class field* $Sh(\phi,\mathfrak{a})$ of the type $(F,\phi,\mathfrak{a})$ is the class field of F' corresponding to $U(\phi,\mathfrak{a})$ defined in 4.17:

$$Sh(\phi,\mathfrak{a}) = (F'^{ab})^{(U(\phi,\mathfrak{a}),F')} = \mathbb{C}^{\widetilde{(U(\phi,\mathfrak{a}),F')}}. \tag{4.20}$$

4.6 Moduli Fields

In section 4.4, Theorem 4.24, we learned that each abelian CM-variety has a projective model defined over a number field. The proof of this theorem is not effective. So we look for a method to find an algebraic definition field in special situations. It should be as small as possible and of easy and effective access, at the same time. The moduli fields introduced by SHIMURA-TANIYAMA [79] are remarkable candidates for this purpose. For their definition we continue the discussion of the functorial application of $\mathrm{Aut}\,\mathbb{C}$ to projective varieties X, see section 4.4.

In this way we can correctly define $\mathcal{X}^{\mu}$ for $\mu \in \mathrm{Aut}\,\mathbb{C}$, $\mathcal{X} = cl(X)$ the class of projective varieties $\mathbb{C}$-isomorphic to X, by representants. We set

$$\mathrm{Stab}\,cl(X) = \{\mu \in \mathrm{Aut}\,\mathbb{C};\, X^{\mu} \cong X\}.$$

Definition 4.29. The fixed field of Stab $cl(X)$ in $\mathbb{C}$ is called the *moduli field* of X (or of $cl(X)$). It is denoted/defined by

$$M(X) = M(cl(X)) = \mathbb{C}^{\mathrm{Stab}\, cl(X)}.$$

We come back now to complex abelian varieties A with complex multiplication $\iota : F \longrightarrow \mathbb{Q} \otimes \mathrm{End}\, A$ of type (F, ϕ). For $\mu \in \mathrm{Aut}\, \mathbb{C}$ the μ-transform $(A, \iota)^\mu = (A^\mu, \iota^\mu)$ is also an abelian CM-variety. Looking at the representations on the tangent spaces it is easy to see that $(A, \iota)^\mu$ is of type $(F, \mu \circ \phi)$. So the type doesn't change if and only if μ belongs to $\mathrm{Aut}\,(\mathbb{C}/F')$ by definition 4.15 of the reflex field F'. Refining the above definitions we set

$$\mathrm{Stab}\, cl(A, \iota) = \{\mu \in \mathrm{Aut}\, \mathbb{C};\ (A, \iota)^\mu \cong (A, \iota)\};$$
$$M(A, \iota) = \mathbb{C}^{\mathrm{Stab}\, cl(A,\iota)}.$$

Lemma 4.30. *For the CM-variety (A, ι) of type (F, ϕ) with reflex field F' and moduli field $M(A, \iota)$ it holds that $F' \subseteq M(A, \iota)$.*

Proof: It suffices to check that $\mathrm{Stab}\, cl(A, \iota) \subseteq \mathrm{Stab}\,(\phi)$. If μ stabilizes $cl(A, \iota)$, then the representations of F in the tangent spaces T_A or T_{A^μ}, respectively, are equivalent. Therefore A^μ has the same type (F, ϕ) as A has, hence $\mu \in \mathrm{Stab}\,(\phi)$, which was to be proved. □

We give now the precise definition of algebraic polarizations. A *polarized abelian variety* (as scheme over $\mathbb{C}$) is a pair $(A, \mathcal{C})$ consisting of an abelian variety $A/\mathbb{C}$ and a $\mathbb{Q}$-line in $\mathbb{Q} \otimes \mathrm{Pic}^a(A)$ containing an ample divisor class; $\mathrm{Pic}^a(A)$ denotes the group of algebraic equivalence classes of divisors on A. We say that $(A, \mathcal{C})$ is defined over k, if and only if $\mathcal{C}$ can be represented by an ample divisor C defined over k such that also the embedding $C \longrightarrow A$ is defined over k. We say also that $\mathcal{C}$ is defined over k in this case. If A is defined over k, then one can find a polarization $\mathcal{C}$ of A also defined over k: One takes first a polarization defined over $\bar{\mathbb{Q}}$, which is in fact defined over a finite GALOIS extension L of k. Let D be a representing divisor and set $C = \sum\limits_{\sigma \in \mathrm{Gal}\,(L/k)} D^\sigma$.

In obvious manner one introduces the μ-transforms $(A, \mathcal{C})^\mu$ for

$$\mu \in \mathrm{Aut}\, \mathbb{C}, cl(A, \mathcal{C})^\mu,\ \mathrm{Stab}\, cl(A, \mathcal{C})$$

and the *moduli fields of polarized abelian* CM-varieties

$$M(A, \mathcal{C}) = M(cl(A, \mathcal{C})) = \mathbb{C}^{\mathrm{Stab\ cl}(A,\mathcal{C})},$$
$$M(A, \iota, \mathcal{C}) = \mathbb{C}^{\mathrm{Stab\ cl}(A,\iota,\mathcal{C})}.$$

Lemma 4.31. *Let k be a definition field of the polarized abelian CM-variety $(A, \iota, \mathcal{C})$. Then it holds that $M(A, \iota, \mathcal{C}) \subseteq k$.*

Proof: (see [79], I.4.2, Proposition 14). For $\mu \in \mathrm{Aut}(\mathbb{C}/k)$ we have an obvious isomorphism $(A, \iota, \mathcal{C})^{\mu} = (A, \iota, \mathcal{C})$, hence $\mu \in \mathrm{Stab}\, cl(A, \iota, \mathcal{C})$. □

Proposition 4.32. *Let (A, ι) be an abelian CM-variety, $\mathcal{C}$ as above. Then there exist algebraic definition fields k of $A, \iota, \mathcal{C}$. For each such field one has the following inclusions:*

$$\begin{array}{c} F' \subseteq M(A, \iota) \subseteq M(A, \iota, \mathcal{C}) \subseteq k \subseteq \overline{\mathbb{Q}}, \\ \cup\!\shortmid \\ M(A, \mathcal{C}) \end{array} \tag{4.21}$$

where F' is the reflex field of the type (F, ϕ) of (A, ι).

Proof: The existence of a small definition field $k \subseteq \overline{\mathbb{Q}}$ has been first verified by SHIMURA-TANIYAMA in [79], see Theorem 4.24. The other inclusions are obvious or come from the Lemmas 4.30 or 4.31, respectively. □

4.7 The Main Theorem of Complex Multiplication

We want to connect the moduli field of a polarized CM-variety $(A, \iota, \mathcal{C})$ of type $(F, \phi, \mathfrak{a})$ with the SHIMURA class field of the same type introduced in 4.5. For this purpose we refine the notion of types again taking into account the polarization. Via projectively embedding theta functions one corresponds to the polarization $\mathcal{C}$ a (unique, up to $\mathbb{Q}$-multiplication), RIEMANN form $E : T_A \times T_A \longrightarrow \mathbb{C}$ ($\mathbb{R}$-bilinear, skew-symmetric, non-degenerate with rational values on $\phi(\mathfrak{a} \times \mathfrak{a})$). It is useful to choose a basic form E of this class. By definition, it takes integral values on $\phi(\mathfrak{a} \times \mathfrak{a})$ and is not an integral multiple of a form of the same kind. With these notations the polarized abelian CM-variety $(A, \iota, \mathcal{C})$ is said to be *of type* $(F, \phi, \mathfrak{a}, E)$. If there is no danger of misunderstanding, then we identify A with its standard torus model $\mathbb{C}^g/\phi(\mathfrak{a})$, see 4.3. Since $F = \mathbb{Q} \otimes \mathfrak{a}$, the embedding ϕ of F into $\mathbb{C}^g$ induces an embedding $F/\mathfrak{a} \longrightarrow A_{tor}$ into the set of torsion points of $A(\mathbb{C})$. We will denote this embedding also by ϕ.

The RIEMANN form E is said to be *ϕ-admissible*, if

$$E(D_{\phi}(f)z, w) = E(z, D_{\phi}(f)w) \text{ for all } f \in F, z, w \in \mathbb{C}^g = T_A.$$

In this case also the polarization $\mathcal{C}$ corresponding to E is called admissible.

From now on we assume that the multiplication field F is a *CM-field*, that means a totally imaginary quadratic extension of a totally real number field. For example, the CM-algebra of a simple abelian CM-variety is in any case a CM-field.

But A needs not to be simple in our applications. Under the CM-field condition there exists an admissible polarization on A (see [48], I. 4, Theorem 4.5). We will work only with polarized abelian CM-varieties of admissible type $(F, \phi, \mathfrak{a}, E)$. The following most important theorem holds for them.

Theorem 4.33 (SHIMURA-TANIYAMA's Main Theorem of Complex Multiplication, see [48] III.6). *With the above assumptions and notations let $\mu \in \mathrm{Aut}(\mathbb{C}/F')$ with restriction $\mu|F'^{ab} = (s, F')$ for a suitable $s \in \mathbb{A}^*_{F'}$. Then it holds that:*

(i) *$(A, \iota, \mathcal{C})^\mu$ is of type $(F, \phi, N'(s^{-1})\mathfrak{a}, \mathbb{N}(s)E)$;*

(ii) *With the componentwise action of the finite part of the idele $N'(s^{-1}) \in \mathbb{A}^*_{F'}$ on $F/\mathfrak{a} = \oplus_p F_p/\mathfrak{a}_p$ the following diagram is commutative:*

$$\begin{array}{ccc} F/\mathfrak{a} & \xrightarrow{\phi} & A_{tor} \\ N'(s^{-1})\Big\downarrow & & \Big\downarrow \mu = (s, F') \\ F/N'(s^{-1})\mathfrak{a} & \longrightarrow & A^\mu_{tor} \end{array} \tag{4.22}$$

As immediate consequence we get the following relation with moduli fields.

Theorem 4.34 (SHIMURA-TANIYAMA [79]; [78], V.5.5; see also [48], V.4). *Let $(A, \iota, \mathcal{C})$ be a polarized abelian variety of admissible CM-type $(F, \phi, \mathfrak{a}, E)$. Then the corresponding SHIMURA class field and moduli field coincide:*

$$Sh(\phi, \mathfrak{a}) = M(A, \iota, \mathcal{C}).$$

Proof: We remember that $Sh = Sh(\phi, \mathfrak{a}) = \mathbb{C}^{(\widetilde{U, F'})}, U = U(\phi, \mathfrak{a})$, see (4.20). For $M = M(A, \iota, \mathcal{C})$ we first show that $M \subseteq Sh$. This follows immediately from

$$(\widetilde{U, F'}) \subseteq \mathrm{Stab}\, cl(A, \iota, \mathcal{C}). \tag{4.23}$$

So we take an automorphism $\mu \in (\widetilde{s, F'})$ for $s \in U$. By the Main Theorem of Complex Multiplication 4.33 (i) the μ-transform $(A, \iota, \mathcal{C})^\mu$ is of type $(F, \phi, N'(s^{-1})\mathfrak{a}, \mathbb{N}(s)E)$. By definition of U in (4.17) there is a $\beta \in F$ such that $N'(s^{-1})\mathfrak{a} = \beta$ and $\mathbb{N}(s^{-1}) = \beta\overline{\beta} \in \mathbb{Q}$. Comparing the standard torus models of $(A, \iota, \mathcal{C})$ and $(A, \iota, \mathcal{C})^\mu$ we get

$$A \cong \mathbb{C}^g/\phi(\mathfrak{a}) \cong \mathbb{C}^g/\phi(\beta\mathfrak{a}) \cong A^\mu, \quad \mathbb{N}(s)E = (\beta\overline{\beta})^{-1} \cdot E \in \mathbb{Q} \cdot E.$$

Therefore $(A, \iota, \mathcal{C}) \cong (A, \iota, \mathcal{C})^\mu$, hence $\mu \in \mathrm{Stab}\, cl(A, \iota, \mathcal{C})$.

Conversely, take $\mu \in \mathrm{Stab}\, cl(A, \iota, \mathcal{C})$; then we have an isomorphism $(A,\iota,\mathcal{C}) \xrightarrow{\sim} (A,\iota,\mathcal{C})^\mu$. On the torsion level it has been made precise multiplying $F/\mathfrak{a}$ by $N'(s^{-1})$,

$\mu \in \widetilde{(s,F')}$, see 4.33 (ii). On the other hand A and A^{μ} have equivalent standard torus models. More precisely, the isomorphism $A \xrightarrow{\sim} A^{\mu}$ corresponds to a β-multiplication $\mathbb{C}^g/\phi(\mathfrak{a}) \longrightarrow \mathbb{C}^g/\phi(\beta\mathfrak{a})$ for a suitable $\beta \in F^{\times}$. But $(A, \iota, \mathcal{C})$ is of type $(F, \phi, N'(s^{-1})\mathfrak{a}, \mathbb{N}(s)E)$ by 4.33 (i). Comparing both presentations we get

$$N'(s^{-1})\mathfrak{a} = \beta\mathfrak{a}, \quad \mathbb{N}(s^{-1}) = \beta\overline{\beta} \in \mathbb{Q}^{\times}.$$

Therefore s belongs to U and $\mu \in \widetilde{(s, F')}$, hence

$$\mathrm{Stab}\,(A, \iota, \mathcal{C}) \subseteq \widetilde{(U, F')} \tag{4.24}$$

in contrast to (4.23). It follows that $Sh \subseteq M$. The identity we were looking for is proved. □

4.8 SHIMURA Class Fields by Special Values

In this section we join the effective construction of transcendent functions th_1, th_2, th_3, th_4 described in Section 3.4 with the arithmetic and algebraic constructions around complex multiplication of abelian varieties. Algebraic geometry and number theory touch each other intimately in Main Theorem 4.33 of Complex Multiplication and its application 4.34 to moduli fields. In the sense of section 4.1 we are going on to prove that

4.35. *The triple* $(\mathbb{B}, \mathbb{B}_{\mathrm{sing}}, th)$ *is a ball model (especially, a solution model) of* HILBERT*'s 12-th Problem.*

We have to give precise definition for the special arguments (singular moduli). Generally, an *arithmetic point* on the SIEGEL half space $\mathbb{H}_g$ is a point of $\mathbb{H}_g(\overline{\mathbb{Q}}) = \mathbb{H}_g \cap \mathbb{G}l_g(\overline{\mathbb{Q}})$. An *arithmetic ball point* (w.r.t. $K, \Gamma = \mathbb{U}((2,1), \mathcal{O}_K)$ or $*$ defined in (3.38)) is a point of

$$\mathbb{B}(\overline{\mathbb{Q}}) = \mathbb{B} \cap \mathbb{H}_3(\overline{\mathbb{Q}}) = *^{-1}(\mathbb{H}_3(\overline{\mathbb{Q}})).$$

A (principally) polarized abelian variety A of dimension g with complex multiplication determines an arithmetic point $\Omega \in \mathbb{H}_g(\overline{\mathbb{Q}})$. Namely, we know from section 4.2 that $A \cong \mathbb{C}^g/\phi(\mathfrak{a})$, $\mathfrak{a}$ a $\mathbb{Z}$-lattice in a number field F. So we have $A \cong \mathbb{C}^g/\Lambda_{\Pi}, \Pi \in \mathbb{M}\mathrm{at}_{g\times 2g}(\overline{\mathbb{Q}})$, Λ_{Π} the $\mathbb{Z}$-lattice generated by the columns of the index matrix. On the other hand A is isomorphic to $\mathbb{C}^g/\Lambda_{(E_g|\Omega)}$. Therefore Π and $(E_g|\Omega)$ represent the same element of the double coset $\mathbb{G}l_g(\mathbb{C})\backslash\mathbb{M}at_{g\times 2g}(\mathbb{C})/\mathbb{G}l_{2g}(\mathbb{Z})$. Hence there exist elements $G \in \mathbb{G}l_g(\mathbb{C})$ and $\Sigma \in \mathbb{G}l_{2g}(\mathbb{Z})$ such that $\Pi\Sigma = (G|G\Omega)$. Now it is clear that G, $G\Omega$ and finally $\Omega = G^{-1}(G\Omega)$ belong to $\mathbb{G}l_g(\bar{\mathbb{Q}})$. In particular, we have at our disposal the following well-known

Lemma 4.36. *If C is a (smooth) curve of genus g and its Jacobian variety $J(C)$ corresponding to $\Omega \in \mathbb{H}_g$ has complex multiplication, then Ω is an arithmetic point of $\mathbb{H}_g$.* □

Definition 4.37. A *singular module* on $\mathbb{B}$ is an isolated fixed point of (an element) of $\mathbb{U}((2,1),K)$.

Definition 4.38. Let $\tau \in \mathbb{B}$ be an arbitrary ball point and $J_\tau = \mathbb{C}^3/\Lambda_{(E_3|*\tau)}$. If J_τ can be decomposed up to isogeny into simple abelian varieties with complex multiplication, then τ is called a *DCM-module* (decomposed CM). If J_τ itself has complex multiplication, then τ is called a *CM-module*; τ is a *simple CM-module (SCM-module)*, if, additionally, J_τ is simple.

Consider the analytic family $\mathbf{J}/\mathbb{B}$ with fibres J_τ at $\tau \in \mathbb{B}$. By the SCHOTTKY-TORELLI diagram (3.24) J_τ coincides up to isomorphy with the Jacobian threefold $J(C_\tau)$ of the PICARD curve C_τ defined in (3.25). The group $\langle\rho\rangle$ of third unit roots acts in obvious manner on C_τ sending (x,y) to $(x,\rho y)$. This action induces a K-multiplication on J_τ, $K = \mathbb{Q}(\sqrt{-3})$. An element g of $\mathbb{U}((2,1),K)$ satisfies its characteristic equation whose degree over K is not greater than 3. We assume for a moment the general case that $K[g]$ is a field extension of degree 3 over K. Also in general, g has a unique, hence an isolated fixed point σ inside of $\mathbb{B}$. Then $K[g]$ acts on the tangent space of J_σ. This action is lattice-compatible. Consequently, $J_\sigma \cong J(C_\sigma)$ has $K[g]$-multiplication. Then J_σ is a CM-variety and σ a CM-module. If we give up the assumption of $K[g]$ to be a field extension of degree 3 over K, then it is also quite elementary but a little bit longer to verify that the isolated fixed point σ, if it exists in $\mathbb{B}$, is a DCM-module. So all singular modules are DCM-modules. Conversely, it is true that all Jacobians of PICARD curves with decomposed complex multiplication can be effectively discovered as (isolated) fixed points of elements of $\mathbb{U}((2,1),K)$. Neglecting a messy but straightforward part of proof we announce the following

Proposition 4.39 (FEUSTEL, unpublished). *The point $\sigma \in \mathbb{B}$ is a singular module if and only if it is a DCM-module.* □

In an earlier paper we proved already the following Theorem 4.40. For CM-modules it is part of the Main Theorem after.

Theorem 4.40 ([34]). *If $\sigma \in \mathbb{B}$ is a singular module, then*

$$th(\sigma) = (th_1(\sigma) : th_2(\sigma) : th_3(\sigma) : th_4(\sigma))$$

is an algebraic point of $\mathbb{P}^2$.

For a CM-module $\sigma \in \mathbb{B}$ we define $(F_\sigma, \phi_\sigma, \mathfrak{a}_\sigma)$ as type of the (polarized) Jacobian $J_\sigma = J(C_\sigma)$ of the PICARD curve C_σ. The reflex field is denoted by F'_σ and the corresponding SHIMURA class field $\mathrm{Sh}\,(\phi_\sigma, \mathfrak{a}_\sigma)$ has been defined in section 4.5. The title result of this section is the following

Main Theorem 4.41. *Let $\sigma \in \mathbb{B}$ be a CM-module. Then we have with the above notations and a suitable subgroup $S_4(\sigma)$ of the symmetric group S_4 a tower of algebraic number fields*

$$F'_\sigma(\mathrm{th}\,(\sigma)) \,/\, F'_\sigma(\mathrm{th}\,(\sigma))^{S_4(\sigma)} \,/\, F'_\sigma/K, \tag{4.25}$$

where the middle extension is abelian; $F'_\sigma(\mathrm{th}\,(\sigma))^{S_4(\sigma)}$ coincides with the SHIMURA *class field $\mathrm{Sh}\,(\phi_\sigma, \mathfrak{a}_\sigma)$, hence with the moduli field $M(J_\sigma, \iota_\sigma, \mathcal{C}_\sigma), \mathcal{C}_\sigma$ the canonical polarization of $J_\sigma = J(C_\sigma)$. The class field extension is unramified, if the following additional "ideal condition" (I) is satisfied:*

(I) $\mathfrak{a}_\sigma$ *is a (fractional) ideal of the relative cubic CM-field of F_σ/K.*

In this case $F'_\sigma(\mathrm{th}\,(\sigma))^{S_4(\sigma)}$ is a subfield of the HILBERT *class field of F'_σ.*

Proof: We first remark that

$$\mathbb{Q}(\mathrm{th}\,(\sigma)) := \mathbb{Q}(\ldots, \mathrm{th}_i(\sigma)/\mathrm{th}_j(\sigma), \ldots)_{i=1}^4, \quad (\mathrm{th}_j(\sigma) \neq 0), \tag{4.26}$$

is a definition field of $cl(C_\sigma)$ and of $cl(J_\sigma)$.

For the curve C_σ this follows immediately from the definition of C_σ in (3.25). Moreover, if k is a definition field of a curve C, then it is also a definition field of its Jacobian variety $J(C)$, more precisely of $cl(J(C))$. For this well-known fact we refer the reader to MILNE's article [53] in [17]. Consequently, $\mathbb{Q}(\mathrm{th}\,(\sigma))$ is a definition field of $cl(J_\sigma)$ or of J_σ itself, without loss of generality. The same is true for $(J_\sigma, \mathcal{C}), \mathcal{C}$ a suitable polarization of J_σ. We recall the identity

$$M(J_\sigma, \iota_\sigma, \mathcal{C}) = (\mathrm{Sh}\,(\phi_\sigma, \mathfrak{a}_\sigma), \tag{4.27}$$

see Theorem 4.34. Next notice that

$$M(J_\sigma) \subseteq M(C_\sigma) = \mathbb{Q}(\mathrm{th}\,(\sigma))^{S'_4(\sigma)}, \quad S'_4(\sigma) \subseteq S_4. \tag{4.28}$$

Proof: Set $S'_4(\sigma) = S_4 \cap (\mathrm{Aut}\,(\mathbb{Q}(\mathrm{th}\,(\sigma)))$. In more precise words the group $S'_4(\sigma)$ consists of all permutations of the four special values $\mathrm{th}\,(\sigma)$ which are extendable to a field automorphism of $\mathbb{Q}(\mathrm{th}\,(\sigma))$. We remember that

$$(\mathrm{th}_1(\sigma) : \mathrm{th}_2(\sigma) : \mathrm{th}_3(\sigma) : \mathrm{th}_4(\sigma))/S_4 \in \mathbb{P}^2/S_4$$

is the moduli point of the PICARD curve C_σ, see (2.12). So for $\mu \in \mathrm{Aut}\,\mathbb{C}$ one has the following equivalent conditions:

$$\begin{aligned}
\mu \in \mathrm{Stab}\,cl(C_\sigma) &\longleftrightarrow C_\sigma^\mu \cong C_\sigma \quad \longleftrightarrow \quad cl(C_\sigma^\mu) = cl(C_\sigma) \\
&\longleftrightarrow (\mathrm{th}_1(\sigma) : \ldots : \mathrm{th}_4(\sigma))^\mu \equiv \mathrm{th}_1(\sigma) : \ldots : \mathrm{th}_4(\sigma)) \bmod S_4 \\
&\longleftrightarrow \mathrm{th}_i(\sigma)^\mu/\mathrm{th}_j(\sigma)^\mu = \mathrm{th}_{\pi(i)}(\sigma)/\mathrm{th}_{\pi(j)}(\sigma) \\
&\qquad \text{for all } i, j \in \{1,2,3,4\} \text{ and a suitable } \pi \in S_4 \\
&\longleftrightarrow \mu \in S'_4(\sigma)
\end{aligned}$$

On the other hand we have $\mathrm{Stab}\,cl(C_\sigma) \subseteq \mathrm{Stab}\,cl(J_\sigma)$, hence $M(J_\sigma) \subseteq M(C_\sigma)$ by the definition of moduli fields. □

Now we are able to prove the essential part of Theorem 4.40.

4.42. The definition field $\mathbb{Q}(\mathrm{th}\,(\sigma))$ of the CM-PICARD curve C_σ, hence also $F'_\sigma(\mathrm{th}\,(\sigma))$, is an algebraic number field.

Proof: Let $\mathcal{C}_\sigma$ be the canonical polarization of J_σ. The polarized Jacobian variety $\mathrm{Jac}\,(C_\sigma) = (J_\sigma, \mathcal{C}_\sigma)$ has an algebraic definition field k by Proposition 4.32. From Lemma 4.31 we know that

$$M(\mathrm{Jac}\,(C_\sigma)) \subseteq k \subset \overline{\mathbb{Q}}.$$

By TORELLI's Theorem the moduli field $M(C_\sigma)$ coincides with $M(\mathrm{Jac}\,(C_\sigma))$. Together with the identity in (4.28) we see that $\mathbb{Q}(\mathrm{th}\,(\sigma))^{S'_4(\sigma)}$ is a number field. Since $\mathbb{Q}(\mathrm{th}\,(\sigma))$ is a finite extension, it is a number field, too. □

Altogether we have the following inclusions:

$$\begin{array}{ccccc} M(J_\sigma,\mathcal{C}) & \subseteq & M(J_\sigma,\iota_\sigma,\mathcal{C}) & = & \mathrm{Sh}(\phi_\sigma,\mathfrak{a}_\sigma) \supseteq F'_\sigma \supseteq K \\ \cup\!\mid & & & & \\ M(J_\sigma) & \subseteq & M(C_\sigma) & = & M(\mathrm{Jac}\,(C_\sigma)) \\ & & & & \| \\ & & & & \mathbb{Q}(\mathrm{th}\,(\sigma))^{S'_4(\sigma)} \subseteq \mathbb{Q}(\mathrm{th}\,(\sigma)) \subset \overline{\mathbb{Q}} \end{array} \tag{4.29}$$

The canonical polarization $\mathcal{C}_\sigma$ is admissible, hence

$$M(J_\sigma,\iota_\sigma,\mathcal{C}) = Sh(\phi_\sigma,\mathfrak{a}_\sigma) = M(J_\sigma,\iota_\sigma,\mathcal{C}_\sigma). \tag{4.30}$$

Together with (4.29) we get the first part of

4.43. For our CM-modules $\sigma \in \mathbb{B}$ and $S_4(\sigma) = S_4 \cap Gal(F'_\sigma(th(\sigma))/F'_\sigma)$ it holds that

$$K \subseteq F'_\sigma \subseteq M(J_\sigma,\iota_\sigma,\mathcal{C}_\sigma) = Sh(\phi_\sigma,\mathfrak{a}_\sigma) = F'_\sigma(th(\sigma))^{S_4(\sigma)} \subseteq F'_\sigma(th(\sigma)) \subset \overline{\mathbb{Q}}.$$

We calculated the moduli field $M(C_\sigma) = M(J_\sigma,\mathcal{C}_\sigma)$, forgetting the complex multiplication, in (4.28). By adjunction of the reflex field F'_σ one obtains the moduli field $M(J_\sigma,\iota_\sigma,\mathcal{C}_\sigma)$. For a proof of this fact we refer to [48], p. 125 or to [78], V.5 5., Proposition 5.17. Altogether we get the following diagram of field extensions:

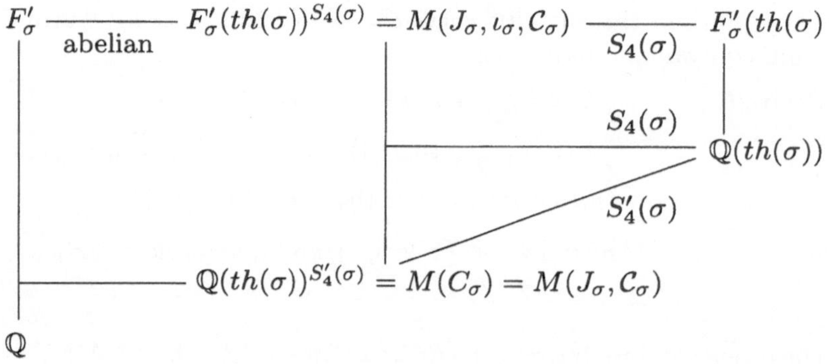

The second identity of 4.43 is proved. □

Now we have proved the first part of Theorem 4.41 because the moduli field $M(J_\sigma, \iota_\sigma, \mathcal{C}_\sigma)$ coincides with the SHIMURA class field $Sh(\phi_\sigma, \mathfrak{a}_\sigma)$ which is, by the definition, an abelian extension of the reflex field F'_σ. It remains to prove that this extension is unramified, if $\mathfrak{a}_\sigma$ is a (fractional) ideal of F_σ. This follows easily from the construction of SHIMURA class fields as we will show now.

Lemma 4.44 (see [48], V.4, Thm. 4.1(ii)). *Let A be an abelian CM-variety of type $(F, \phi, \mathfrak{a})$ such that $\mathfrak{a}$ is a fractional ideal of F. Then the abelian extension $Sh(\phi, \mathfrak{a})/F'$ is unramified.*

For the convenience of the reader we repeat the short proof going back to the construction of the SHIMURA class field $Sh(\phi, \mathfrak{a})$ in section 4.5. Via reciprocity it corresponds to the idele group $U(\phi, \mathfrak{a})$ defined in (4.17). It suffices to verify that $U(\phi, \mathfrak{a})$ contains the whole unit group $\mathcal{O}^*_{F'}$ of $\mathcal{O}_{F'}$. This is a well-known necessary and sufficient criterion for the corresponding class field to be unramified (see e.g. [57]). So let ε be a unit of F'. Then, with the notation of (4.17), also $N'(\varepsilon) \in F$ is a unit. Since $\mathfrak{a}$ is a fractional ideal it holds that

$$N'(\varepsilon)\mathfrak{a} = \mathfrak{a} = 1 \cdot \mathfrak{a}.$$

The relations of the right-hand side of (4.17) are satisfied for $s = \varepsilon$, $\beta = 1$. Hence ε belongs to $U(\phi, \mathfrak{a})$. The lemma is proved, and at the same time we finish the proof of Theorem 4.41. For the relation with the HILBERT class field we refer to Definition 4.1. □

4.9 Special Points on SHIMURA Varieties of $\mathbb{U}(2,1)$

We would like to translate the main result of this chapter to the modern language of SHIMURA varieties. In a more general context this has been done by DELIGNE [20]. We restrict ourselves to establish a dictionary for PICARD modular surfaces. For proofs we refer to [19], [20] and [62]. We start with the definition of polarized HODGE structures and with the characterization of polarized abelian varieties in these terms. Forget for a moment the complex structure of the tangent spaces of abelian varieties of dimension g. As real vector spaces all of them are isomorphic. The complex structure is defined by the non-central action of an element I, $I^2 = -1$. It extends to an action of the group $\mathbb{C}^\times$ in obvious manner. With a glance to holomorphic and antiholomorphic differential forms $(H^{1,0}, H^{0,1})$ it is convenient to consider also antiholomorphic actions. For this purpose $\mathbb{C}^\times$ is understood as group $\mathbb{S}(\mathbb{R})$ of real points of the real algebraic group $\mathbb{S}$ well-defined by reduction of ground field:
$\mathbb{S} = R_{\mathbb{C}/\mathbb{R}}\mathbb{G}_m$ where $\mathbb{G}_m$ is the multiplicative group:

$$\mathbb{G}_m(A) = A^* \quad \text{(units of } A)$$

for commutative rings A with unit 1, especially $\mathbb{G}_m(\mathbb{C}) = \mathbb{C}^\times$. The inclusion $\mathbb{R}^\times \hookrightarrow \mathbb{C}^\times$ and the norm map $\mathbb{C}^\times \longrightarrow \mathbb{R}^\times$ are defined by algebraic group morphisms (at real points)

$$w : \mathbb{G}_{m\mathbb{R}} \longrightarrow \mathbb{S},\ t : \mathbb{S} \longrightarrow \mathbb{G}_{m\mathbb{R}},$$

respectively.

4.45. *Let V be a real vector space of finite dimension. It is equivalent to give one of the following three kinds of data*

(i) *a representation $\rho : \mathbb{S} \longrightarrow \mathbb{G}l(V)$ of weight $n \in \mathbb{Z}$, that means:*

$$\rho w(x)v = x^n v \text{ for } x \in \mathbb{R}^\times, v \in V;$$

(ii) *a* HODGE *bigraduation of weight n of $V_\mathbb{C} = V \otimes_\mathbb{R} \mathbb{C}$, that means:*

$$V_\mathbb{C} = \otimes_{p+q=n} V^{p,q} \text{ such that } \overline{V}^{p,q} = V^{q,p};$$

(iii) *a* HODGE *filtration of weight n of $V_\mathbb{C}$, that means a descending filtration $\{F^p\}$ such that $F^p(V_\mathbb{C}) \oplus \overline{F}^{n-p+1}(V_\mathbb{C}) \xrightarrow{\sim} V_\mathbb{C}$.*

Relating the data one has $F^p(V_\mathbb{C}) = \oplus_{i \geq p} V^{i,j}$ and, with obvious notations, $zv^{p,q} = z^p \overline{z}^q v^{p,q}$. The WEIL *operator* $I = \rho$ (i) acts on $V^{p,q}$ by $Iv^{p,q} = i^{p-q} v^{p,q}$.

Definition 4.46. A HODGE *structure H of weight n* is a torsion free $\mathbb{Z}$-module $H_\mathbb{Z}$ of finite rank together with a HODGE bigraduation on $H_\mathbb{C} = H_\mathbb{Z} \otimes \mathbb{C}$ of weight n. A *polarization of a* HODGE *structure* H of weight n is a morphism (of HODGE structures) $\psi : H \times H \longrightarrow \mathbb{Z}(-n)$ such that the real bilinear form $(2\pi i)^n \psi(x, Iy)$ on $H_\mathbb{R} = H_\mathbb{Z} \otimes \mathbb{R}$ is symmetric and positive definite.

Tacitly we introduced above the HODGE-TATE *module* $\mathbb{Z}(n)$ as special HODGE structure of weight n (type $(-n, -n)$) supported by $\mathbb{Z}(n)_\mathbb{Z} = (2\pi i)^n \mathbb{Z}$, $\mathbb{Z}(n)_\mathbb{C} = \mathbb{C}$. The *type of a* HODGE *structure* consists of all pairs (p, q) such that $V^{p,q} \neq 0$.

Theorem 4.47 (RIEMANN). *The functor $A \mapsto H_1(A, \mathbb{Z})$ gives an equivalence between the category of polarized abelian varieties and the category of polarized* HODGE *structures (of weight one) of type $\{(0,1), (1,0)\}$.*

We have already seen that the action of certain algebraic groups play an important role. We remember to special symmetric spaces as the ball or the SIEGEL half spaces and to the groups of holomorphic automorphisms on them (unitary groups, symplectic groups). We get more specification if we try to extend the action of $\mathbb{S}$ on V to the action of a bigger algebraic group. If the group is defined over a number field, say $\mathbb{Q}$, then we have p-adic groups in analogy to abelian varieties over number fields. The adelic points of groups allow to define "special points" with a glance to the torsion points of abelian varieties.
Let $\mathbb{G}$ be a real linear algebraic group, $I \in \mathbb{G}(\mathbb{R})$, with central square, V a real

representation of $\mathbb{G}$ of finite dimension. If there exists on V a $\mathbb{G}$-invariant bilinear form ψ such that $\psi(x, Iy)$ is symmetric and positive definite on $V \times V$, then V is called *I-polarizable* and ψ an *I-polarization.* This property of I depends for given $\mathbb{G}$, V only on the $\mathbb{G}(\mathbb{R})$-conjugation class of I.
Now we restrict ourselves to reductive algebraic groups $\mathbb{G}$ defined over $\mathbb{Q}$. Our earlier $\mathbb{S}$-actions will be carefully extended by means of commutative diagrams

$$\begin{array}{ccccc} & w & & t & \\ \mathbb{G}_{m\mathbb{R}} & \longrightarrow & \mathbb{S} & \longrightarrow & \mathbb{G}_{m\mathbb{R}} \\ \| & & \downarrow h & & \| \\ \mathbb{G}_{m\mathbb{R}} & \longrightarrow & \mathbb{G}_{\mathbb{R}} & \longrightarrow & \mathbb{G}_{m\mathbb{R}} \\ & w & & t & \end{array} \tag{4.31}$$

where $tw(x) = x^{-2}$ at real points in the second row in analogy to the first row defined earlier. Up to h, all morphisms of the diagram are considered to be fixed.

A finite dimensional representation $V_{\mathbb{Q}}$ of $\mathbb{G}$ defined over $\mathbb{Q}$ is *homogeneous of weight* n, if $w(x)v = x^n v$ for $v \in V, x \in \mathbb{R}^{\times}$. Such $V_{\mathbb{Q}}$ is endowed with a *rational* HODGE *structure* of weight n by means of h in diagram (4.31). A *polarization* (of weight n) of the $\mathbb{G}$-representation V is a $\mathbb{G}$-invariant form $\psi : V_{\mathbb{Q}} \times V_{\mathbb{Q}} \longrightarrow \mathbb{Q}(n)$ such that $\psi(x, h(i)y)$ defines a positive definite symmetric form on $V_{\mathbb{R}}$. The $\mathbb{G}$-invariance implies $\mathbb{S}$-invariance via pull-back along h, see diagram (4.31). Therefore ψ is a polarization of the rational HODGE structure $V_{\mathbb{Q}}$. It depends only on the $\mathbb{G}(\mathbb{R})$-conjugacy class of h such that ψ is a polarization of V. Moreover, the representation $V_{\mathbb{Q}}$ is polarizable if and only if the representation $V_{\mathbb{R}}$ is $h(i)$-polarizable.

Now we are able to understand a dual rational homogeneous variant of RIEMANN's theorem 4.47. Let A be an abelian variety over $\mathbb{C}$. Instead of $H^1(A, \mathbb{Q})$ we consider its dual $H_1(A, \mathbb{Q}) = H$ together with polarizations of type $\{(-1,0)(0,-1)\}$. Let $[\,,\,]$ be a skew-symmetric $\mathbb{R}$-bilinear form on $H = H_{\mathbb{Q}} \otimes \mathbb{R}$. The group $\mathbb{G}p(H)$ of *symplectic similitudes* is defined by

$$\mathbb{G}p(H)(\mathbb{R}) = \{g \in \mathbb{G}l(H)(\mathbb{R}); [gx, gy] = \alpha_g \cdot [x,y], x, y \in H, \alpha_g \in \mathbb{R}^{\times} \text{ suitable}\}. \tag{4.32}$$

Consider quadruples $(V_{\mathbb{Q}}, \psi, V_{\mathbb{Z}}, h)$, where $V_{\mathbb{Q}}$ is a vector space over $\mathbb{Q}, \psi$ a non-degenerate skew-symmetric form on $V, V_{\mathbb{Z}}$ a lattice in V and $h : \mathbb{S} \longrightarrow \mathbb{G}p(V)$ a homomorhism as in (4.31) such that $\psi(x, h(i)x)$ is positive on $V \backslash 0$. Remember that h defines a HODGE structure on $V_{\mathbb{C}}$ with filtration $F(h)$.

Theorem 4.48 (RIEMANN). *There is an equivalence of categories induced by the (mutually inverse) correspondences of objects*

$$\begin{aligned} (A, p) &\mapsto (H_1(A, \mathbb{Q}), \psi, H_1(A, \mathbb{Z}), h), \\ (V_{\mathbb{Q}}, \psi, V_{\mathbb{Z}}, h) &\mapsto (F(h)^0 \backslash V_{\mathbb{Q}} \otimes \mathbb{C}/V_{\mathbb{Z}}, p) \end{aligned} \tag{4.33}$$

with (algebraically) polarized complex abelian varieties (A, p) *and quadruples as defined above.*

Our main object in this section comes from PICARD modular groups. Let K be an arbitrary imaginary quadratic number field and $\langle\,,\,\rangle : K^3 \times K^3 \longrightarrow K$ a hermitian form of signature $(2,1)$. For simplicity we assume that $\langle\,,\,\rangle$ is unimodular, represented by the diagonal matrix $\mathrm{diag}(1,1,-1)$, say. We define a $\mathbb{Q}$-vector space $V = V_{\mathbb{Q}}$ by $V(\mathbb{Q}) = K^3$ and a skew-symmetric bilinear form

$$[\,,\,] : V \times V \longrightarrow \mathbb{Q}, \quad (v,w) \mapsto Tr_{K/\mathbb{Q}}\langle v, \bar{\delta} w\rangle,$$

$\delta \in K^\times \cap \mathbb{R}_+ i$, which is uniquely determined up to a positive rational factor. The group $\mathbb{G}$ of symplectic similitudes of $(V, [\,,\,])$ is defined over $\mathbb{Q}$. For a number field L we denote by L^* the algebraic $\mathbb{Q}$-group $R_{L/\mathbb{Q}}\mathbb{G}_{m,L}$. Then it is easy to see that K^* appears as center of $\mathbb{G}$.

We get back well-known classical groups as

$$\mathbb{G}'(\mathbb{Q}) = \mathbb{SU}((2,1),K), \;(\mathbb{G}' \text{ the derived group}),$$
$$(\mathbb{G}/K^*)(\mathbb{Q}) = \mathbb{PU}((2,1),K), \;\mathbb{G}(\mathbb{Q}) = \mathbb{GU}((2,1),K)$$

and the (special) PICARD modular group $\Gamma = \mathbb{G}'(\mathbb{Z}) = \mathbb{SU}((2,1),\mathcal{O}_K)$. Moreover, the correspondence $\nu : g \mapsto \alpha_g \cdot \det^{-1} g$ yields an exact sequence

$$1 \longrightarrow \mathbb{G}' \longrightarrow \mathbb{G} \xrightarrow{\nu} K^* \longrightarrow 1\,. \tag{4.34}$$

We define a HODGE structure h by

$$h_0 : \mathbb{S} \longrightarrow \mathbb{G},\; z \mapsto \begin{pmatrix} \bar{z} & 0 & 0 \\ 0 & \bar{z} & 0 \\ 0 & 0 & z \end{pmatrix}^{-1}. \tag{4.35}$$

By composition with the inner automorphism of $\mathbb{G}_{\mathbb{R}}$ we get the $\mathbb{G}(\mathbb{R})$-conjugacy class X of h_0:

$$X = \{gh_0g^{-1} := (g\Box g^{-1}) \circ h_0;\, g \in \mathbb{G}(\mathbb{R})\}. \tag{4.36}$$

Then $[.,h(i).]$ is symmetric and positive definite on $V_{\mathbb{R}}$. The conjugacy class X is uniquely determined by the hermitian form $\langle\,,\,\rangle$ in this way. We have

$$V_h = V_h^{-1,0} = \{v \in V_{\mathbb{C}};\, h(z)v = z^{-1}v,\, z \in \mathbb{S}(\mathbb{R})\} \tag{4.37}$$

and the $[.,h(i).]$-orthogonal splittings

$$V_h^+ = V_{\mathbb{R}} \cap \bar{V}_h\,,\; V_h^- = V_{\mathbb{R}} \cap V_h \tag{4.38}$$

of $V_{\mathbb{R}}$ in positive and negative subspaces.

As we know, our groups $\mathbb{G}(\mathbb{R}), \mathbb{G}'(\mathbb{R}) \ldots$ act on the two-ball $\mathbb{B}$. They act also transitively on X by definition. Comparing isotropy groups we discover $\mathbb{B}$

from the HODGE structures introduced above. More precisely, we have an analytic isomorphism

$$X \xrightarrow{\sim} \mathbb{B}, \quad h := gh_0g^{-1} \mapsto g(0), \quad g \in \mathbb{U}((2,1),\mathbb{C}). \tag{4.39}$$

In this way the ball points are interpreted as morphisms h. The corresponding vectors $\mathfrak{a}$ in $\mathbb{C}^3$ ($\mathbb{P}\mathfrak{a} \in \mathbb{B}$) come from V_h^-.

Observe that we have more structure on our $V_\mathbb{Q}$ or $V_\mathbb{R}$, namely a multiplication with elements of K. Denote the form $[\,,\,]$ by ψ. For $k \in K$ it holds that

$$\psi(kv, w) = \psi(v, \bar{k}w) \text{ on } V_\mathbb{R}. \tag{4.40}$$

We extend our quadruples of Theorem 4.48 above now to quintuples $(V_\mathbb{Q}, \psi, V_\mathbb{Z}, h, \kappa)$, κ a K-multiplication on V with property (4.40) and look for the corresponding objects in the sense of RIEMANN's Theorem. We find polarized abelian varieties with compatible K-multiplication: $(A, p, \iota), \iota : K \longrightarrow \mathbb{Q} \otimes \operatorname{End} A$ such that the involution of complex conjugation on K is induced by the polarization $p : A \longrightarrow A^*, A^* = \operatorname{Hom}(A, \mathbb{G}_{a\mathbb{C}})$ the dual abelian variety of A. This means that the conjugation coincides with the corresponding ROSATI involution (transposition) on $\mathbb{Q} \otimes \operatorname{End} A : \iota(\bar{k}) = \iota(k)^*, \ k \in K$. The correspondence constructions can be extended to F-multiplications, F/K a finite (CM)-extension or even to L-multiplications, where L is a semisimple $\mathbb{Q}$-algebra with involution acting faithfully on abelian varieties.

An adelic extension of RIEMANN's Theorem will give a nice interpretation of ball quotients $\Gamma'\backslash\mathbb{B}$ by PICARD modular groups Γ' as moduli spaces. Remember that the local rings $\mathbb{Z}_p$ are open compact in $\mathbb{Q}_p$, $p \in \operatorname{Spec}\mathbb{Z}$. So open compact subgroups of $\mathbb{G}(\mathbb{A}^f)$, $\mathbb{A} = \mathbb{A}_\mathbb{Q}$ the ring of $\mathbb{Q}$-adeles, $\mathbb{A}^f = \prod_p \mathbb{Q}_p$ the finite adeles, play an important role.

The n-torsion points of an abelian variety A are denoted by A_n. The TATE *module* $\hat{T}(A)$ of A is defined by

$$\hat{T}(A) = \varprojlim A_n = \hat{\mathbb{Z}} \otimes H_1(A, \mathbb{Z}) = \prod_{l \in \operatorname{Spec}\mathbb{Z}} T_l(A)$$

$$\hat{Z} = \varprojlim \mathbb{Z}/n\mathbb{Z} = \prod_p \mathbb{Z}_p, \quad T_l(A) = \varprojlim \mathbb{Z}/l^n\mathbb{Z}.$$

We also need

$$\hat{V}(A) = \mathbb{A}^f \otimes_{\hat{\mathbb{Z}}} \hat{T}(A) = \mathbb{A}^f \otimes H_1(A, \mathbb{Q})$$

and the *semi-trace map*

$$t : L \longrightarrow \bar{\mathbb{Q}}, l \mapsto Tr(l; F(h_0)^0\backslash V_\mathbb{C}).$$

Theorem 4.49 ([20], 4.11). *Let $\mathbb{G}$ be the algebraic $\mathbb{Q}$-group of symplectic similitudes of a symplectic space $(V_{\mathbb{Q}}, \psi)$, ψ skew-symmetric, non-degenerate, with compatible L-multiplication, that means $\psi(lv, w) = \psi(v, l^* w)$, and $\mathbf{K}$ an open compact subgroup of $\mathbb{G}(\mathbb{A}^f)$. There is a quasiprojective $\mathbb{C}$-variety ${}_{\mathbf{K}}Sh_{\mathbb{C}}(\mathbb{G}, h_0)$ whose points correspond bijectively to the isogeny classes of quadruples $(A, \iota, \bar{p}, \bar{\eta})$ consisting of:*

(i) *a complex abelian variety A;*
(ii) *an L-multiplication $\iota : L \longrightarrow \mathbb{Q} \otimes \mathrm{End}\, A$ such that $Tr(\iota(l), \mathrm{Lie}\, A) = t(l), l \in L$, for fixed semi-trace t;*
(iii) *a (homogeneous) polarization class $\bar{p}$ inducing the involution on L;*
(iv) *a class $\bar{\eta} = \eta \bmod \mathbf{K}$ of L-linear symplectic similitudes $\eta : \hat{V}(\mathbb{A}) \xrightarrow{\sim} V \otimes \mathbb{A}^f$;*

such that the following conditions are satisfied:

(a) *The polarized L-module $(H_1(A, \mathbb{Q}), \psi_p)$ is isomorphic to (V, ψ);*
(b) *under the isomorphism the HODGE structure h of $H_1(A)$ coincides up to conjugation with a given fixed HODGE structure $h_0 : \mathbb{S} \longrightarrow \mathbb{G}_{\mathbb{R}}$.*

The rough moduli space of the the objects of the Theorem is the double coset

$$ {}_{\mathbf{K}}Sh_{\mathbb{C}} = {}_{\mathbf{K}}Sh_{\mathbb{C}}(\mathbb{G}, h_0) = \mathbf{K}_\infty \times \mathbf{K} \backslash \mathbb{G}(\mathbb{A}) / \mathbb{G}(\mathbb{Q}). \tag{4.41} $$

Sending $\mathbf{K}$ to $g\mathbf{K}g^{-1}$, $g \in \mathbb{G}(\mathbb{A}^f)$, one obtains a projective system $\{{}_{\mathbf{K}}Sh_{\mathbb{C}} \longrightarrow {}_{g\mathbf{K}g^{-1}} Sh_{\mathbb{C}}\}$. The projective limit

$$ Sh_{\mathbb{C}} = \lim_{\leftarrow \mathbf{K}} Sh_{\mathbb{C}} \tag{4.42} $$

is called the SHIMURA *variety* (of $(\mathbb{G}, h_0)$ over $\mathbb{C}$). Sometimes one calls also ${}_{\mathbf{K}}Sh_{\mathbb{C}}$ a SHIMURA variety (of *level* $\mathbf{K}$).

We go back to the PICARD modular case around (4.34). For this special case with K-multiplication we have the following

Corollary 4.50. *The variety ${}_{\mathbf{K}}Sh = {}_{\mathbf{K}}Sh(\mathbb{G}, h_0)$ is the rough moduli scheme of the functor ${}_{\mathbf{K}}M$ corresponding to each connected Noetherian K-scheme S the set of isogeny classes of quadruples $(A, \iota, \bar{p}, \bar{\eta})$, where*

(i) *A/S is a projective abelian scheme over S of relative dimension 3;*
(ii) *$\iota : K \longrightarrow \mathbb{Q} \otimes \mathrm{End}\, A$ a homomorphism with invertible $\mathcal{O}_S$-module $\{a \in \mathrm{Lie}\, A; \iota(k)a = ka$ for all $k \in K\}$ (trace condition);*
(iii) *$\bar{p} = p \circ \iota(\mathbb{Q}^\times)$ with a polarization $p : A \longrightarrow A^*$ such that $\iota(k)^* \circ p = p \circ \iota(\bar{k})$ for all $k \in K$;*
(iv) *$\bar{\eta}$ corresponds to a geometric point s of S a $\pi_1(S, s)$-invariant class of $K \otimes_{\mathbb{Q}} \mathbb{A}^f$-module isomorphisms (modulo $\mathbf{K}$), $\eta : \hat{V}(A_s) \longrightarrow V \otimes_{\mathbb{Q}} \mathbb{A}^f$, such that there exists $\beta \in \mathbb{A}^f(1)_s$ and $p \in \bar{p}$ producing the RIEMANN form of p_s as $\beta \circ \psi \circ (\eta \times \eta) : \hat{V}(A_s) \times \hat{V}(A_s) \longrightarrow \mathbb{A}^f(1)_s$.*

We look for small subfields E of $\mathbb{C}$ over which our SHIMURA varieties can be defined. More precisely, let $M = M_E$ be a variety defined over the field E such that $M_{\mathbb{C}} \cong Sh_{\mathbb{C}}$ (over $\mathbb{C}$), then M is called a *model* (over E) of the SHIMURA variety Sh. It is also denoted by $Sh_E(\mathbb{G}, h)$ for $Sh_{\mathbb{C}} = Sh_{\mathbb{C}}(\mathbb{G}, h)$. We will define below a special definition field $E(\mathbb{G}, h)$ for reductive algebraic $\mathbb{Q}$-groups $\mathbb{G}$. First we present its meaning in the case of our symplectic groups.

We observed already in (4.39), (4.36) that the HODGE structure morphisms $h : S \longrightarrow G$ appear as "points" of symmetric domain. Intuitively we consider them as "points" of the corresponding SHIMURA variety $Sh(\mathbb{G}, h_0)$. More precisely, we look for (de)compositions

$$h : \mathbb{S} \xrightarrow{h'} H_{\mathbb{R}} \xrightarrow{u_{\mathbb{R}}} \mathbb{G}_{\mathbb{R}} \tag{4.43}$$

with algebraic subtorus $u : H \hookrightarrow \mathbb{G}$ such that $h = u \circ h'$ is $\mathbb{G}(\mathbb{R})$-conjugate to h_0.

Theorem 4.51. *Let $\mathbb{G}/\mathbb{Q}$ be one of our groups of symplectic type, $h = u \circ h'$ a "point" as described in* (4.43), *$E = E(\mathbb{G}, h_0)$ or $E(H, h')$ the corresponding definition fields. Then $Sh_{\mathbb{C}}(\mathbb{G}, h_0)$ has a model $Sh_E = Sh_E(\mathbb{G}, h_0)$ defined over E, and the canonical morphism*

$$Sh_{\mathbb{C}}(H, h') \longrightarrow Sh_{\mathbb{C}}(\mathbb{G}, h_0)$$

is defined over the composite $E(H, h') \cdot E$.

The E-model in the theorem is uniquely defined up to E-isomorphy. It is called the *canonical model* of the SHIMURA variety $Sh(\mathbb{G}, h_0)$. For proofs we refer to [20] or to [52].

These field extensions $E(H, h')E/E$ are most interesting in class field theory. This will be clarified by the following version of the Main Theorem of Complex Multiplication. We specify the situation of Theorem 4.49 to monogen L-modules $V_{\mathbb{Q}}$. The algebraic closure of $E = E(\mathbb{G}, h_0)$ is denoted by $\bar{E}$. Consider the isogeny classes of quadruples $(A, \iota, \eta, \bar{p})$ of objects described in 4.49 (i), ..., (iv), additionally: A is defined over $\bar{E}; t(L) \subseteq E$.

For $L = F$, $[F : \mathbb{Q}] = 2 \cdot \dim A$ we have to do with abelian varieties with complex multiplication; E appears as reflex field.

Theorem 4.52 (SHIMURA-TANIYAMA [20]). *The* GALOIS *group $Gal(\bar{E}/E)$ acts on the above classes of quadruples via its greatest abelian quotient. Let $\varphi : E^*(\mathbb{A}^f) \longrightarrow$ $\mathrm{Gal}\,(E^{ab}/E)$ be the canonical map of abelian class field theory. Together with the reciprocity law $r = r(\mathbb{G}, h) : E^* \longrightarrow \mathbf{T}$ one gets for $e \in E^*(\mathbb{A})$ with finite component $e^f \in E^*(\mathbb{A}^f)$ the following description of the* GALOIS *group action:*

$$\varphi(e)(A, \iota, \eta, \bar{p}) = (A, \iota, r(e^f) \cdot \eta, \bar{p}).$$

We have to explain definition field and reciprocity law. For this purpose we complexify h and pull it back to $\mathbb{G}_m$ along the morphism $r : \mathbb{G}_m \longrightarrow \mathbb{S}$ defined as follows: With obvious notations the compositions

$$\mathbb{G}_m \xrightarrow{r} \mathbb{S} \xrightarrow{z^p\bar{z}^q} \mathbb{G}_m,\ p+q=n,$$

coincide with $x \mapsto x^p$, respectively, at complex points. Then $\mathbb{G}(\mathbb{C})$ acts via conjugation on the set of modified (complex) HODGE morphisms

$$\mu_h = h \circ r : \mathbb{G}_m \xrightarrow{r} \mathbb{S}_\mathbb{C} \xrightarrow{h} \mathbb{G}_\mathbb{C}. \tag{4.44}$$

The *field of definition* $E(\mathbb{G}, h)$ of $(\mathbb{G}, h)$ is the definition field of the $\mathbb{G}(\mathbb{C})$-conjugation class of $h \circ r$ in (4.44). This is the fixed field of all $\sigma \in \operatorname{Aut}\mathbb{C}$ such that $\sigma_G \circ h \circ r$ is conjugate to $h \circ r$, $\sigma_G : \mathbb{G}_\mathbb{C} \xrightarrow{\sim} \mathbb{G}_\mathbb{C}$ induced by σ at complex points.

In order to explain the reciprocity morphism we take into account a further composition. As in (4.34) we have an exact sequence

$$0 \longrightarrow \mathbb{G}' \longrightarrow \mathbb{G} \xrightarrow{\nu} \mathbf{T} \longrightarrow 0,$$

$\mathbf{T}$ a torus. The composed morphism

$$\mu = \mu_h \circ \nu = \nu \circ r \circ h : \mathbb{G}_m \xrightarrow{r} \mathbb{S} \xrightarrow{h} \mathbb{G} \xrightarrow{\nu} \mathbf{T} \tag{4.45}$$

depends only on the conjugation class of $h \circ r$. Therefore it is defined over $E = E(\mathbb{G}, h)$. If we apply the functor $R_{E/\mathbb{Q}}$ of WEIL reduction of ground field and decompose by the norm map, then we obtain the *reciprocity morphism*

$$r(\mathbb{G}, h) : E^* = R_{E/\mathbb{Q}}(\mathbb{G}_{mE}) \overset{R_{E/\mathbb{Q}}(\mu^{-1})}{\longrightarrow} R_{E/\mathbb{Q}}(\mathbf{T}_E) \overset{N_{E/\mathbb{Q}}}{\longrightarrow} \mathbf{T} \tag{4.46}$$

we are looking for.

For the rest of this section we restrict ourselves to the PICARD modular case. We fix the notations introduced around (4.34). The morphism μ_h in (4.44) can be explicitly described as follows:

$$\begin{aligned} \mathbb{G}_{m\mathbb{C}} \longrightarrow G_{m\mathbb{C}} \times G_{m\mathbb{C}} \xrightarrow{\sim} \mathbb{S}_\mathbb{C} \xrightarrow{h} \mathbb{G}_\mathbb{C} \\ z \longmapsto (z,1),\ (x,y) \mapsto (xy, \bar{x}y). \end{aligned} \tag{4.47}$$

Especially for $h = h_0$ one gets easily the correspondence

$$\mu_{h_0} : \mathbb{G}_{m\mathbb{C}} \longrightarrow G_{m\mathbb{C}},\quad z \mapsto \begin{pmatrix} (1,z) & 0 & 0 \\ 0 & (1,z) & 0 \\ 0 & 0 & (z,1) \end{pmatrix}^{-1}.$$

Now it is obvious that the definition field is $E(\mathbb{G}, h) = K$.

The combination with ν of (4.34) yields $\mu : \mathbb{G}_{mK} \longrightarrow (K^*)_K$ and finally the PICARD specialization of (4.46)

$$r(\mathbb{G}, h_0) : K^* \longrightarrow (K \otimes_\mathbb{Q} K)^* \longrightarrow K^* \tag{4.48}$$

$$k \longmapsto k^{-1}\bar{k}^{-1}. \tag{4.49}$$

Main Example 4.53. Let $K = \mathbb{Q}(\sqrt{-3})$ be the field of EISENSTEIN numbers and $V_{\mathbb{Z}} = \mathcal{O}_K^3$. We have a canonical (maximal) open compact subgroup

$$\mathbf{K} = \{g \in \mathbb{G}(\mathbb{A}^f);\ \ g\hat{V}_{\mathbb{Z}} = \hat{V}_{\mathbb{Z}}\}$$

of $G(\mathbb{A}^f)$.

Set $\delta = i\sqrt{3} \in \mathbb{R}_+ i$. According to our earlier notations we have a perfect pairing

$$\psi = [\,,\,] : V_{\mathbb{Z}} \times V_{\mathbb{Z}} \longrightarrow \mathbb{Z}.$$

Since $K^*\nu(\mathbf{K}) = K^*(\mathbb{A}^f)$ the corresponding SHIMURA variety ${}_{\mathbf{K}}Sh$ is geometrically connected and

$${}_{\mathbf{K}}Sh(\mathbb{C}) = \Gamma\backslash\mathbb{B}, \Gamma = \mathbb{PU}((2,1), \mathcal{O}_K).$$

This simplification works for all imaginary quadratic number fields K with class number 1 and with $\delta \in \mathbb{R}_+ i$ generating the difference of the extension $K/\mathbb{Q}$. The corresponding moduli problem simplifies to:

${}_{\mathbf{K}}Sh$ is the rough moduli scheme of the functor corresponding to each connected Noetherian K-scheme S the set of isomorphy classes of triples (A, ι, p) consisting of

(i) A projective abelian scheme over S;
(ii) An $\mathcal{O}_K$-multiplication $\iota : \mathcal{O}_K \hookrightarrow \operatorname{End} A$ such that

$$\{a \in \operatorname{Lie} A; \iota(\omega)a = \omega a \ \text{ for all } \ \omega \in \mathcal{O}_K\}$$

is an invertible $\mathcal{O}_S$-module;
(iii) p is a principal polarization of A, compatible with the $\mathcal{O}_K$-multiplication $(\iota(\omega)^* \circ p = p \circ \iota(\bar{\omega}))$, such that for the geometric points $s \in S$ there is an $\mathcal{O}_K \otimes_{\mathbb{Z}} \hat{\mathbb{Z}}$-isomorphism $\eta : \hat{T}(A_s) \xrightarrow{\sim} \hat{V}_{\mathbb{Z}}$ and $\beta \in \hat{Z}(1)_s$ representing the RIEMANN form of p_s as $\beta \circ \psi \circ (\eta \times \eta) : \hat{T}(A_s) \longrightarrow \hat{Z}(1)_s$.

Remark 4.54. In the case of EISENSTEIN numbers we have the same moduli space $\Gamma\backslash\mathbb{B}$ as for PICARD curves. The Jacobian threefolds of PICARD curves fit into the above functor of principally polarized abelian threefolds with K-multiplication. Since the moduli spaces coincide we found a precise characterization of Jacobians of PICARD curves in these K-multiplication terms.

Now it is interesting to ask for DCM-points and the class field extensions defined by them. DELIGNE's interpretation of "points" is described in (4.43). We need points which allow higher than K-multiplication for the corresponding abelian varieties. Then we can apply the version 4.52 of SHIMURA-TANIYAMA's theorem, if the multiplication algebra L is great enough ($V_{\mathbb{Q}}$ is L-monogen).

Definition 4.55. If H and u in (4.43) are defined over $\mathbb{Q}$, then $H \subseteq G, u, h', h$ and also the points $(h, \mathbf{K}g) \in {}_{\mathbf{K}}Sh(\mathbb{C})$ are called *special.*

We are well-prepared to find the image torus H for any point $h : \mathbb{S} \longrightarrow \mathbb{G}$. Namely, in (4.38) we defined an orthogonal splitting $V_{\mathbb{R}} = V_h^- \oplus V_h^+$. On the subspace V_h^- we have an $\mathbb{S}$-($\mathbb{S}(\mathbb{R}) = \mathbb{C}^\times$)-multiplication $v \mapsto zv, z \in \mathbb{C}$, see (4.37) and take inverses. So V_h^- is a complex(ified) vector space of dimension 1. On V_h^+ we have an $\mathbb{S}$-multiplication $v \mapsto \bar{z}v,\ z \in \mathbb{C}$. The complex dimension is 2. Altogether we get a splitting

$$h : \mathbb{S} \stackrel{(id,\bar{id},\bar{id})}{\longrightarrow} \mathbb{S} \times \mathbb{S} \times \mathbb{S} \longrightarrow \mathbb{G}\,. \tag{4.50}$$

We call the $\mathbb{S}$-multiplication on V_h^- an id-multiplication, that on V_h^+ an $\bar{id}$- or $(\bar{id}, \bar{id})$-multiplication and that on $V_{\mathbb{R}}$ the $(id, \bar{id}, \bar{id})$-multiplication induced by h. The morphisms of (4.50) are defined over $\mathbb{Q}$, if the first morphism is lifted from $K^* \longrightarrow H_{\mathbb{Q}}$. There are three cases corresponding to (decomposed) complex multiplication:

$$H_{\mathbb{Q}} = \begin{cases} F^*, & F \text{ a CM-field}, \quad [F:K] = 3 \\ F^* \times K^*, & F \text{ a CM-field}, \quad [F:K] = 2 \\ F^* \times F^* \times F^*, & F = K. \end{cases} \tag{4.51}$$

The K-multiplication on our abelian varieties is induced by restriction of h or $(id, \bar{id}, \bar{id})$ to K^*. The triple is called the *type of K-multiplication.* The type of F-multiplication must be an extension of a subtype of K-multiplication.

Lemma 4.56. *There are only three kinds of decomposed complex multiplication of abelian threefolds A of* PICARD *type, namely*

(i) *A is simple with F-multiplicator, $[F : K] = 3$.*

(ii) *$A \approx D \times D \times E$, E an elliptic curve with K-multiplication, D an elliptic curve with L-multiplication, L an imaginary quadratic number field different from K,*

$$F = K \cdot L\,.$$

(iii) *$A \approx E \times E \times E$, E as in (ii).*

Proof: Each isogeny component of A of multiplicity 1 has K-multiplication. If C is such a one-dimensional component, then it has to be isogeneous to E.

Next we exclude that A has a simple abelian surface component B. Denote the multiplication field of B by F.

Then we have a biquadratic field tower

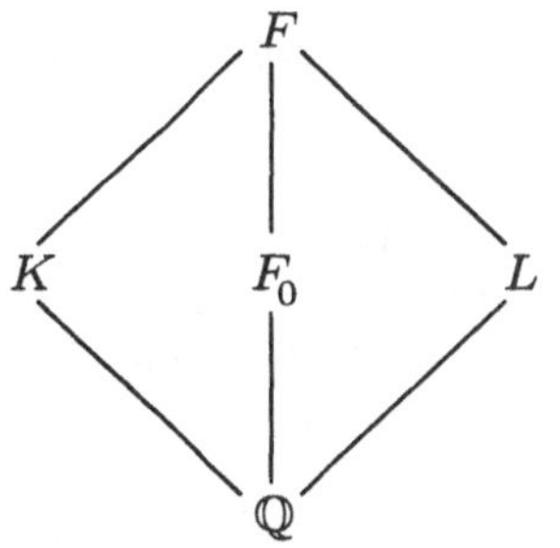

F_0 a real quadratic number field. The type of F-multiplication is denoted by $(\varphi_1, \varphi_2), \varphi_i : F \hookrightarrow \mathbb{C}$. We have some freedom to fix φ_1, but in accordance with the restriction to K-multiplication of subtype $(id, \bar{id})$ or $(\bar{id}, \bar{id})$. In the first case we chose $\varphi_1 = id_F$, in the second $\varphi_1 = (\bar{id})_F$. In any case φ_2 has a fixed subfield. This subfield is different from F_0 because $\varphi_2 \neq \bar{\varphi}_1$ by definition of types. The fixed fields are L or K in the first or second case, respectively. Therefore the type (φ_1, φ_2) is not irreducible; it is lifted from L or K, respectively. This means that B has an isogeny decomposition $D \times D$ or $E \times E$, respectively, D, E as above.

We see that A is decomposed into three elliptic curves, if it is not simple. Since each factor of multiplicity 1 has K-multiplication, there are no other possibilities than those described in (ii) and (iii). □

Starting from h we look for the field $F = F(h)$, especially in the second case of (4.51). Then we have a precise correspondence between the three cases of Lemma 4.56 and (4.51), respectively. The simpler cases (i) and (iii) are left to the reader.

Consider the id-multiplication part V_h^- of $V_{\mathbb{R}}, V = V_{\mathbb{Q}}(\mathbb{Q}) = K^3$. It has complex dimension 1 (the real points with id-complexification). Let V' be the smallest K-linear subspace of V such that $V^- \subseteq V'_{\mathbb{R}}$ (omit the index h). We only investigate in detail the two-dimensional case $\dim_K V' = 2$. Beside of the natural K-multiplication we have a K-multiplication κ on V' of type $(id, \bar{id})$ induced by h (restricted to K^*). Set

$$F = \{f \in \mathrm{End}_K V'; f \circ \kappa = \kappa \circ f\}. \tag{4.52}$$

The projections of $V'_{\mathbb{R}}$ on V^- or on its complement $V^+ \cap V'_{\mathbb{R}}$ are not K-defined because V^- isn't. Otherwise we had $V'_{\mathbb{R}} = V^-$ in contradiction to $\dim_K V' = 2$. Therefore the non-zero elements of F do not kill V^- and we get a field embedding $\varphi_1 : F \hookrightarrow \mathrm{End}_{\mathbb{C}} V^- = \mathbb{C}$.

We use the skew-symmetric $\mathbb{R}$-bilinear form $[\,,\,]$ (defined after 4.48) on $V'_{\mathbb{R}}$ to introduce the $\mathbb{Q}$-linear map $* : F \hookrightarrow \mathrm{End}_K V'$ by the identity $[f^*v, w] = [v, fw]$ on $V' \times V'$. From

$$[f^*\kappa^*v, w] = [\kappa^*v, fw] = [v, \kappa fw] == [v, fkw] = [f^*v, kw] = [k^*f^*v, w]$$

we get $\kappa^* f^* = f^* \kappa^*$, hence, multiplying with κ from left and right, $f^* \kappa = \kappa f^*$. Therefore f^* belongs to F by definition. We show that $*$ is an involution of F:

$$[v, fw] = [f^* v, w] = -[v, f^* w] = -[f^{**} w, v] = [v, f^{**} w].$$

Finally we check that F is a CM-field and $*$ the complex conjugation. Obviously, F is a field extension of K, and $*$ is the extension of the complex multiplication on K. Therefore F is not a real subfield of $\mathbb{C}$. It suffices to show that the $*$-norms $f^* f$ are positive numbers for $f \in F^{\times}$. Remember that $(.,.) = [., \kappa.]$ is symmetric and positive definite on $V_{\mathbb{R}}$, hence on V'. We deduce $f^* f > 0$ from

$$0 < (fv, fv) = [fv, \kappa f v] = [fv, f \kappa v] = [f^* f v, \kappa v] = (f^* f v, v), \ v \neq 0 .$$

The torus morphism h splits into h', h'' described now:

$$\begin{array}{ccc} \mathbb{S} & \xrightarrow{h'} & \mathbb{S} \times \mathbb{S} \subset \mathbb{G}l_K(V')_{\mathbb{R}} \\ \downarrow & & \downarrow \\ K^* & \longrightarrow & F^* \end{array} \qquad \begin{array}{ccc} \mathbb{S} & \xrightarrow{h''} & \mathbb{S} \cong \mathbb{G}l_K(V'')_{\mathbb{R}} \\ \downarrow & & \downarrow \\ K^* & \longrightarrow & K^* \end{array} \tag{4.53}$$

where V'' is the $\langle , \rangle$-orthogonal complement of V' in V, which is K-defined. The $\mathbb{S}$-multiplications h', h'' are of type $(id, \bar{id})$ or of type $(\bar{id})$, respectively. The vertical morphisms are base changes from $\mathbb{Q}$ to $\mathbb{R}$.

Corresponding diagrams can be established also for the first and third case of (4.51) and Lemma 4.56. This is left to the reader.

Now we change back to the very classical terminology characterizing special points of the ball $\mathbb{B}$ with respect to a fixed imaginary quadratic number field K. We write

$$x = (x_1, x_2) \in \mathbb{B}: \ |x_1|^2 + |x_2|^2 < 1;$$

$K(x) = K(x_1, x_2)$ the *coordinate field* of x or $\tilde{x}$ (over K);

$$\tilde{x} = (x_1, x_2, 1), \ \mathbb{P}\tilde{x} = x \in \mathbb{B} = \mathbb{P}\tilde{\mathbb{B}}, < \tilde{x}, \tilde{x} >= |x_1|^2 + J|x_2|^2 - 1 < 0.$$

We say that a set $\sum$ of field embeddings $M \hookrightarrow \mathbb{C}$ *acts orthogonally* at x, if $\sigma(\tilde{x}) = \tilde{x}$ or $< \sigma(\tilde{x}), \tilde{x} >= 0$ for all $\sigma \in \sum$.

Proposition 4.57. *For a special ball point x the definition field F coincides with the coordinate field $K(x)$, the multiplication field defined in* (4.52) *and the toroidal field defined in* (4.51).

The following conditions (1), (2) are necessary and sufficient for $x \in \mathbb{B}$ to be special:

(1) *CM-field condition: $\sigma(\bar{x}) = \overline{\sigma(x)}$ for all $\sigma \in \mathrm{Gal}\,(\bar{\mathbb{Q}}/\mathbb{Q})$;*
(2) *Orthogonality-condition: $\Sigma(F/K) = \{\sigma : F \hookrightarrow \mathbb{C}; \sigma|K = id_K\}$ acts orthogonally at x.*

Remark 4.58. Corresponding to the three cases (i), (ii), (iii) of Lemma 4.56, see also (4.51), the latter condition can also be rewritten as:

(2′) There exists an $<,>$-orthogonal basis $\tilde{x}, \tilde{y}, \tilde{z}$ of $\mathbb{C}^3$ of one of the following types:

(2.i) $(\tilde{x}, \tilde{y}, \tilde{z}) = (\sigma_1(\tilde{x}), \sigma_2(\tilde{x}), \sigma_3(\tilde{x}))$, $\sigma_1, \sigma_2, \sigma_3$ the three different embeddings of F into $\mathbb{C}$ over K;

(2.ii) $(\tilde{x}, \tilde{y}, \tilde{z}) = (\sigma_1(\tilde{x}), \sigma_2(\tilde{x}), \tilde{z})$, σ_1, σ_2 the two different embeddings of F into $\mathbb{C}$ over K, $\tilde{z} \in K^3$ generating the orthogonal complement of $\mathbb{C}\tilde{x} + \mathbb{C}\tilde{y}$ in $\mathbb{C}^3$;

(2.iii) $\tilde{x}, \tilde{y}, \tilde{z} \in K^3$, $\tilde{y}, \tilde{z}$ an (arbitrary) orthogonal basis of the orthogonal complement of $\tilde{x}$ in $\mathbb{C}^3$ with coordinates in K.

If $\phi = (\tau_1, \tau_2, \tau_3)$ or (τ_1, τ_2), $\tau_1 = \mathrm{id}$, is the corresponding CM-type in case (i) or (ii), respectively, then it holds that $\tau_2 = \bar{\sigma}_2$, $\tau_3 = \bar{\sigma}_3$ (up to a change of the (2,3)-numeration in case (i)).

Proof: We only discuss the case (ii). Here V' is two-dimensional, its orthogonal complement V'' in $V = K^3$ is K-defined. Hence it is generated by a vector $\tilde{z} \in K^3$. $V'_{\mathbb{R}}$ splits orthogonally into V^- and V_2. We know that F defined in (4.52) acts on V^- and $f^* = \bar{f}$ for $f \in F$. Therefore the action of F on V' is orthogonal. Namely, if $\langle v, w \rangle = 0$, then

$$\langle fv, fw \rangle = \langle f^* fv, w \rangle = \langle \bar{f} fv, w \rangle = \bar{f} f \langle v, w \rangle = 0.$$

Thus the action of F on V' splits into characters $\rho_i \in \mathrm{End}(V_i)$, $v_i \mapsto \varphi_i(f)v_i$, $V_1 = V^-$, $f \in F$, $v_i \in V_i$.

We prove that $F = K(x)$ for $\tilde{x}$ generating $V_1 = V^-$. Let σ be a field automorphism of $\mathbb{C}$. Then the following conditions are equivalent:

$$\sigma(x) = x, \quad \sigma(\tilde{x}) = \tilde{x}, \quad \sigma V^- = V, \quad \sigma|V^- = \mathrm{id}, \quad \sigma\varphi_1 = \varphi_1, \quad \sigma|F = \mathrm{id}_F.$$

Therefore $F = K(x)$. The condition (1) is satisfied because F is a CM-field. The smallest K-subspace V' of V with $V^- \subseteq V_{\mathbb{R}}$ is obviously generated by $\tilde{x}$ and $\mu(\tilde{x})$, $\mu \in \mathrm{Gal}(\bar{\mathbb{Q}}/K)$, $\mu|F \neq \mathrm{id}_F$. The F-action on V^- extends uniquely to a K-defined action on V' sending $a\tilde{x} + b\mu(\tilde{x})$ to $af\tilde{x} + b\mu(f)\mu(\tilde{x})$, $f \in F$. The splitting of the F-action in different characters is unique. Therefore, $\mu(\tilde{x})$ generates V_2 and (2) follows immediately.

Now assume conversely, that the conditions (1), (2) are satisfied for $x \in \mathbb{B}$. Then the set of all $\sigma(\tilde{x})$, $\sigma \in \mathrm{Aut}_K(\mathbb{C})$, form an orthogonal system of non-zero vectors of $\mathbb{C}^3$. The maximal possible number of elements of such a system is 3. Therefore x is algebraic and $[K(x) : K] \leq 3$. Now set $V^- = \mathbb{C}\tilde{x}$, $V' = \sum_{\sigma \in \mathrm{Gal}(\bar{\mathbb{Q}}/K)} \sigma(V^-)$. Then $V'_{\mathbb{R}}$ has a K-model V'. Set $F = K(x)$ and define the

F-multiplication on the σ-component of $V'_{\mathbb{R}}$ by multiplication with $\sigma(f)$, $f \in F$. By (1) we know that F is a CM-field. The hermitian form $\langle , \rangle$ is negative definite on the id-component V^- and positive definite on the other components because of the signature type (2,1). Now it is not difficult to see that the field F comes out as the endomorphism algebra defined in (4.52). The rest is clear. □

For further developments we refer to: R.P. LANGLANDS, D. RAMAKRISHNAN (ED.), *The zeta functions of* PICARD *modular surfaces*, Publ. CRM, Montréal, 1992.

5 Transcendental Values of Picard Modular Theta Constants

Introduction

It was G. WUESTHOLZ who refered me during a visit in Berlin to the possibility to prove that the PICARD modular theta covering $\mathrm{th} = (\mathrm{th}_1 : \mathrm{th}_2 : \mathrm{th}_3 : \mathrm{th}_4) : \mathbb{B} \longrightarrow \mathbb{P}^2$ sends non-singular arithmetic points τ of $\mathbb{B}$ to *transcendental points* $\mathrm{th}(\tau)$, which means that $\mathrm{th}(\tau) \notin \mathbb{P}^2(\bar{\mathbb{Q}})$. This has been announced in [34]. Parallel to our investigations of algebraic value a transcendence proof has been carried out by H. SHIGA in [75], [76]. In the meantime SHIGA generalized the transcendence proof to several other cases and to the higher-dimensional SIEGEL coverings of moduli spaces of polarized abelian varieties. The initiating role of our relative situation of Jacobian threefolds of PICARD curves motivated us to present SHIGA's first special proof in a modified form of a lecture in all details. If the PICARD modular situation is understood then it is not difficult to understand the more general results in [76].

5.1 Transcendence at Non-Singular Simple Algebraic Moduli

We first present a modified version of SHIGA's proof of the following theorem. It prepares its extension to all algebraic ball points which are non-singular. Extending Definition 4.38 we call $\tau \in \mathbb{B}$ a *module of simple type*, if the Jacobian $J_\tau \cong J(C_\tau)$ of the PICARD curve C_τ is a simple abelian variety.

Theorem 5.1 ([75]). *Let $\tau \in \mathbb{B}(\bar{\mathbb{Q}})$ be a non-singular module of simple type. Then $\mathrm{th}(\tau) \in \mathbb{P}^2$ is transcendental.*

The fundament of the proof is WUESTHOLZ's Theorem 5.5 below. Before we connect it with PICARD curves we need some simple preliminary facts.

5.2. *Let A be a simple complex abelian variety and G a non-trivial connected algebraic subgroup of $A \times A$, this means that $G \neq 0,\ 0 \times A,\ A \times 0,\ A \times A$. Then it holds that*

$$(i) \quad \dim\ G = \dim\ A = g,$$

$$(ii) \quad p_i(G) = A, \quad i = 1, 2,$$

where $p_i : A \times A \longrightarrow A$ is the projection to the i-th factor.

Proof: The second statement follows immediately from the conditions. In order to prove (i) we consider the following commutative diagram (5.1.3) with exact diagonals which fix our notations:

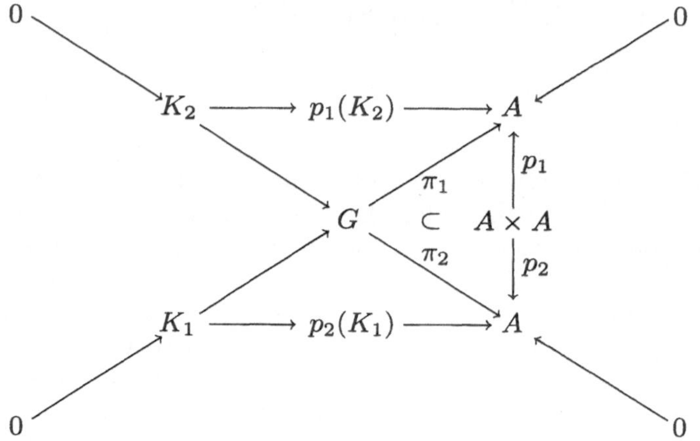

(5.1)

For the proof of (i) we have to check that $\dim K_i = 0$ for $i = 1, 2$. Assume for example that $\dim K_1 > 0$. Then it follows that $\dim p_2(K_1) > 0$. Namely, otherwise we had $p_2(K_1^0) = p_1(K_1^0) = 0$ for the connected component K_1^0 of 0 in K_1, hence

$$K_1^0 \subseteq p_1^{-1}(0) \cap p_2^{-1}(0) = (0 \times A) \cap (A \times 0) = 0.$$

Therefore K_1 is finite, $\dim K_1 = 0$ in contradiction to our assumption. From $\dim p_2(K_1) > 0$ we get $p_2(K_1) = A$ because A is simple, hence $\dim K_1 \geq g$,

$$\dim G = \dim K_1 + \dim A \geq 2g = \dim A \times A,$$

$G = A \times A$, in contradiction to the assumptions of 5.2. Thus it cannot happen that $\dim K_1 > 0$. A glance to the defining diagram (5.1) shows that *(i)* is satisfied. □

5.3. *Under the conditions of* 5.2 *the subgroup* G *of* $A \times A$ *defines an element* $\varphi = \varphi_G$ *of* $\mathbb{Q} \otimes \operatorname{End} A$.

By (ii) the morphism π_1 is an isogeny. Take an opposite isogeny $\pi' : A \to G$. We define ρ_G to be the image of the composition $\pi_2 \circ \pi' : A \to A$ in $\mathbb{Q} \otimes \operatorname{End} A$. Since the endomorphism algebras of abelian varieties are invariants of isogeny classes, the correspondence $G \mapsto \varphi_G$ is well-defined.

We set $T_A = \operatorname{Lie} A$. The exponential map $\exp : T_A \to A$ is a well-defined homomorphism of abelian Lie groups (see e.g. [55], ch. I). It defines an exact sequence

$$0 \longrightarrow \Lambda_A \longrightarrow T_A \xrightarrow{\exp} A \longrightarrow 0$$

with a lattice $\Lambda_A \cong \mathbb{Z}^{2g}$ in T_A. Identify $T_{A \times A}$ with $T_A \oplus T_A$. An endomorphism φ of the abelian variety A defines an abelian subvariety $Graph(\varphi)$ of $A \times A$. Its tangent space $graph(\varphi) = \operatorname{Lie} Graph(\varphi)$ is a g-dimensional linear subspace of $T_A \oplus T_A$. The definition extends to elements φ of $\mathbb{Q} \otimes \operatorname{End} A$ in an obvious manner. With the notations of 5.3 check that

5.4. *φ_G belongs to $\mathbb{Q} \otimes \mathrm{id}_A$ if there exist natural numbers m, n such that*

$$T_G = \{(z, w) \in T_A \oplus T_A; mz = nw\}.$$

□

Theorem 5.5 (WUESTHOLZ [90]). *Let G, G' be connected commutative algebraic groups of positive dimensions defined over $\bar{\mathbb{Q}}$ and $h : G' \to G$ an analytic homomorphism defined over $\bar{\mathbb{Q}}$ (w.r.t. the coefficients of Taylor series of the functions defining the analytic map h at $1_{G'}$). Assume that $h(G')(\bar{\mathbb{Q}}) \neq 0$. Then there is an algebraic subgroup H of G defined over $\bar{\mathbb{Q}}$ such that $H \subseteq h(G')$ and $\dim H > 0$.*

Remark 5.6 ([90]). *If H is maximal with the above property, then it holds that $H(\bar{\mathbb{Q}}) = h(G')(\mathbb{C}) \cap G(\bar{\mathbb{Q}})$.*

We will apply the theorem to Jacobian varieties $J(C)$ of smooth complete algebraic curves C defined over $\bar{\mathbb{Q}}$, more precisely to biproducts $G = J(C) \times J(C)$ and the corresponding exponential map will play the role of h in WUESTHOLZ's theorem. As an intermediate step we apply Theorem 5.5 to the more special situation of abelian varieties A over $\bar{\mathbb{Q}}$ and to the exponential map $\exp : T_A \to A$. Notice that T_A has a $\bar{\mathbb{Q}}$-structure defined by a $\bar{\mathbb{Q}}$-basis of the space $H^0(A, \Omega_A^1)$ of differential forms via duality.

Corollary 5.7 (see SHIGA [75]). *Let A be an abelian variety defined over $\bar{\mathbb{Q}}$, W a linear $\bar{\mathbb{Q}}$-subspace of T_A and $w \neq 0$ an element of W such that $\exp(w) \in A(\bar{\mathbb{Q}})$, then there is a non-finite algebraic subgroup H of A defined over $\bar{\mathbb{Q}}$ and a commutative diagram*

$$\begin{array}{ccc} T_A & \xrightarrow{\exp} & A \\ \uparrow & & \cup\!\!| \\ | & & \exp(W) \\ | & & \cup\!\!| \\ w \in T_H & \longrightarrow & H \end{array} \tag{5.2}$$

Proof: First we check that $\exp(W)(\bar{\mathbb{Q}}) \neq 0$. It may happen that $\exp(w) = 0$, and this is really the case we will deal with. But then we can find a natural number n such that $\exp(w') \neq 0$ for $w' = w/n \in W \backslash 0$. Since $\exp(w')$ is a torsion point of A it lies in $A(\bar{\mathbb{Q}})$. Now the conditions of Theorem 5.5 are satisfied and we obtain the diagram above. We choose H maximal. Then by Remark 5.6 we have

$$\exp(w) \in A(\bar{\mathbb{Q}}) \cap \exp(W)(\mathbb{C}) = H(\bar{\mathbb{Q}}),$$
$$w \in T_H = (\exp^{-1}(H))^0 \subseteq (\exp^{-1}(\exp(W)))^0 = W,$$

where ^{0} denotes the connected component containing $0 \in T_A$. □

Now we consider smooth PICARD curves $C : Y^3 = \prod_{i=1}^4 (X - t_i)$. Remember that a typical basis of $H_1(C, \mathbb{Z})$ is a normal basis of type $*(\alpha_1, \alpha_2, \alpha_3)$, $\alpha_i \in H_1(C, \mathbb{Z})$, where $*$ is defined as

$$* : H_1(C, \mathbb{Z})^3 \longrightarrow H_1(C, \mathbb{Z})^6,$$
$$* : \vec{\alpha} = (\alpha_1, \alpha_2, \alpha_3) \longmapsto (\alpha_1, \alpha_2, -g(\alpha_1); \alpha_3, g(\alpha_2), g^2(\alpha_3)),$$

where $g \in \operatorname{Aut} C$ sends (x, y) to $(x, \delta y)$. We refer to the Definition (2.18) and to diagram (2.43).

We identify a choice of a typical basis with an isometry of symplectic modules $(\mathbb{Z}^6, I) \xrightarrow{\sharp} (H_1(C, \mathbb{Z}), E)$ sending the canonical basis of $\mathbb{Z}^6$ to a typical basis, where E is the intersection product of cycles and I is the skew-symmetric bilinear form represented by $\begin{pmatrix} 0 & E_3 \\ -E_3 & 0 \end{pmatrix}$. The existence of at least one typical basis on each smooth PICARD curve has been proved with Proposition 2.17.

The Jacobian variety of C can be identified with

$$J(C) = H^0(C, \Omega_C)^\vee \backslash H_1(C, \mathbb{Z}),$$

where $\vee$ denotes the dual vector space. We know that

5.8. *There is an explicit bijective correspondence*

$$Iso \backslash \left\{ \begin{array}{ll} (C, (\mathbb{Z}, I) \xrightarrow{\#} H_1(C, \mathbb{Z})); & C \text{ smooth Picard curve} \\ & \# \text{ a typical basis on } C \end{array} \right\} \longleftrightarrow \mathbb{B} \backslash \bigoplus$$

This is nothing else but an intrinsic interpretation of the SCHOTTKY-TORELLI diagram (3.24), see also (2.63). The correspondence from the left to the right is given by period matrices:

$$(C, *\vec{\alpha}) \longmapsto \Pi(C, *\vec{\alpha}) = \int_{*\vec{\alpha}} \vec{\omega} = \begin{pmatrix} *\mathfrak{a} \\ *\mathfrak{b} \\ \overline{*\mathfrak{e}} \end{pmatrix} \longmapsto \mathbb{P}\mathfrak{a} \in \mathbb{B}$$

where $\vec{\omega} =^t (dx/y, dx/y^2, xdx/y^2)$ is the typical basis of $H^0(C, \Omega_C)$,

$$*(A_1, A_2, A_3) = (A_1, A_2, -\rho A_1; A_3, \rho A_2, \bar{\rho} A_3).$$

The period matrix can be written as

$$\Pi(C, *\vec{\alpha}) = (\vec{a}_1, \vec{a}_2, -D \cdot \vec{a}_1; \vec{a}_3, D \cdot \vec{a}_2, \bar{D} \vec{a}_3), \tag{5.3}$$

where D denotes the diagonal matrix $D = \mathrm{diag}(\rho, \bar{\rho}, \bar{\rho})$. Splitting period matrices $\Pi = (\Pi_1|\Pi_2)$ into two quadratic parts we get the period point $\Omega = \Omega(C, *\vec{\alpha}) = \Pi_1^{-1} \cdot \Pi_2 = (\tau_{ij})$ in $\mathbb{H}_3$. From (5.3) we get the matrix relation

$$(\vec{a}_3, D\vec{a}_2, \bar{D}\vec{a}_3) = (\vec{a}_1, \vec{a}_2, -D\vec{a}_1)\Omega. \tag{5.4}$$

More explicitly we write

$$\Pi(C, *\vec{\alpha}) = \begin{pmatrix} A_1, & A_2, & -\rho A_1; & A_3, & \rho A_2, & \bar{\rho} A_3 \\ \bar{B}_1, & \bar{B}_2, & -\bar{\rho}\bar{B}_1; & \bar{B}_3, & \bar{\rho}\bar{B}_2, & \rho\bar{B}_3 \\ \bar{C}_1, & \bar{C}_2, & -\bar{\rho}\bar{C}_1; & \bar{C}_3, & \bar{\rho}\bar{C}_2, & \rho\bar{C}_3 \end{pmatrix}. \tag{5.5}$$

The corresponding period point has been correctly calculated by PICARD, see (2.64).

$$\Omega = \begin{pmatrix} (\bar{\rho}u^2 + 2\rho v)/(1-\bar{\rho}), & -\bar{\rho}u, & (u^2 - \rho v)/(1-\bar{\rho}) \\ -\bar{\rho}u, & \bar{\rho}, & -u \\ (u^2 - \rho v)/(1-\bar{\rho}), & -u, & (\rho u^2 + 2\rho v)/(1-\bar{\rho}) \end{pmatrix} \tag{5.6}$$

with $u = A_2/A_1$, $v = \bar{\rho}A_3/A_1$ (A_1 cannot be equal to 0), and

$$2\,\mathrm{Re}(v) + |u|^2 < 0. \tag{5.7}$$

This is the explicit description of the ball embedding $\mathbb{B} \to \mathbb{H}_3$.

We can describe also the K-multiplication $\iota : K \to \mathbb{Q} \otimes \mathrm{End}\, J(C)$ in an explicit manner by means of (5.3). We have

$$J(C) = \mathbb{C}^3/\Lambda(C, *\vec{\alpha}), \tag{5.8}$$
$$\Lambda(C, *\vec{\alpha}) = \mathbb{Z}\vec{a}_1 + \mathbb{Z}\vec{a}_2 + \mathbb{Z}D\vec{a}_1 + \mathbb{Z}\vec{a}_3 + \mathbb{Z}D\vec{a}_2 + \mathbb{Z}\bar{D}\vec{a}_3,$$

in canonical coordinates. The diagonal matrix $D \in \mathrm{Gl}_3(\mathbb{C})$ represents an isomorphism $\iota(\rho)$ of $J(C)$. This representation extends to K sending $k \in K$ to

$$\iota(k) = \mathrm{diag}(k, \bar{k}, \bar{k}) \in \mathbb{Q} \otimes \mathrm{End}\, J(C)\,. \tag{5.9}$$

5.9. *If the endomorphism algebra of the Jacobian of a* PICARD *curve is simple, then it is either a cubic field extension F of K or isomorphic to K.*

Proof: Use the table of types of division algebras $\mathbb{Q} \otimes \mathrm{End}\, A$, A a simple abelian variety of dimension g over $\mathbb{C}$. It can be found in [55], IV.21:

type	e	d	relation	η
I	e_0	1	$e \mid g$	1
II	e_0	2	$2e \mid g$	$\frac{3}{4}$
III	e_0	2	$2e \mid g$	$\frac{1}{4}$
IV	$2e_0$	d	$e_0 d^2 \mid g$	$\frac{1}{2}$

with the following entries. Denote the center of $\mathbb{Q}\otimes \mathrm{End}\, A$ by Z. This is a (number) field of absolute degree $e = [Z : \mathbb{Q}]$ and $\mathbb{Q} \otimes \mathrm{End}\, A$ is a Z-algebra of Z-dimension d^2. If Z_0 denotes the fixed subfield of Z of the ROSATI involution, then Z_0 is a totally real number field of absolute degree $e_0 = [Z_0 : \mathbb{Q}]$. The types II and III differ by an invariant η, which we do not need and explain.

We work with $A = J(C)$, $g = 3$. Since g is odd the types II, III are killed by the relations in the table. Also type I is impossible because it leads to the contradiction

$$K \subseteq \mathbb{Q} \otimes \mathrm{End}\, J(C) = Z = Z_0 \quad \text{total real.}$$

So we have type IV and the relation of the table yields $d = 1$, hence $Z = \mathbb{Q} \otimes \mathrm{End}\, J(C)$. The cases $e_0 = 1$, $e_0 = 3$ lead to $e = 2$, $e = 6$, hence $Z = K$ or $Z = F$, respectively. □

Lemma 5.10. *Let C be a smooth* PICARD *curve, defined over $\bar{\mathbb{Q}}$, $J(C)$ simple, $\Omega(C, *\vec{\alpha}) = \Omega(u, v) \in \mathbb{H}_3(\bar{\mathbb{Q}})$. If $\mathbb{Q} \otimes \mathrm{End}\, J(C) = K$, then $u \in K$.*

Proof: Assume that $u \notin K$. By Lemma 5.9 it suffices to construct an endomorphism φ of $A = J(C)$ not belonging to $\iota(K)$. Consider in $T_A \oplus T_A$ the linear subspace

$$W : \begin{pmatrix} u & 0 & 0 \\ 0 & 0 & 0 \\ 0 & 0 & 0 \end{pmatrix} \vec{x} = \begin{pmatrix} 1 & 0 & 0 \\ 0 & 0 & 0 \\ 0 & 0 & 0 \end{pmatrix} \cdot \vec{y}\,. \tag{5.10}$$

Since $\Omega(u, v) \in \mathbb{H}_3(\bar{\mathbb{Q}})$ by assumption and $\Omega(u, v)$ has the form described in (5.6) we have $u \in \bar{\mathbb{Q}}$, hence W is defined over $\bar{\mathbb{Q}}$. The pair $(\vec{x}, \vec{y}) = (\vec{a}_1, \vec{a}_2)$ belongs to W because of $u = A_2/A_1$, (5.3) and (5.5). Moreover, $\vec{a}_1, \vec{a}_2 \in \Lambda(C, *\vec{\alpha})$, hence $\exp(\vec{a}_1, \vec{a}_2) = 0$ in $A \times A$. With Corollary 5.7 we obtain a non-trivial algebraic subgroup H of $A \times A$ defined over $\bar{\mathbb{Q}}$, $H \subseteq \exp(W)$. It defines an element φ_H of $\mathbb{Q} \otimes \mathrm{End}\, J(C)$ by 5.2 and 5.3. By construction, see diagram (5.2), we know that $\mathrm{graph}(\varphi_H) \subseteq W$. On the other hand we have

$$\mathrm{graph}(\iota(k)) : \mathrm{diag}(k, \bar{k}, \bar{k})\vec{x} = \vec{y}, \quad k \in K. \tag{5.11}$$

It is not contained in W because of our assumption $u \notin K$. Thus $\varphi_H \notin \iota(K)$, which was to be proved. □

Proof of Theorem 5.1: We first reformulate the theorem. Let $C = C_\tau : Y^3 = \prod_{i=1}^4 (X - \mathrm{th}_i(\tau))$ be the PICARD curve belonging to τ. C is a smooth PICARD curve because $J(C)$ is assumed to be simple. The period point $\Omega = *\tau$ belongs to $\mathbb{H}_3(\bar{\mathbb{Q}})$ because the ball coordinates u, v are algebraic and Ω is given by (5.6). Assume that $\mathrm{th}(\tau)$ is algebraic. Then C is defined over $\bar{\mathbb{Q}}$. If we can prove that $u, v \in K$, then we have a contradiction because $J(C)$ is not simple in this case (see Lemma 5.15 below) in contrast to the assumption of Theorem 5.1. Therefore it suffices to prove the following

Corollary 5.11 of Lemma 5.10. *Under the assumptions of* 5.10 *it holds that* $u, v \in K$, *if* $\mathbb{Q} \otimes \operatorname{End} J(C) = K$.

Proof: The vector $\vec{a}_3$ is a $\bar{\mathbb{Q}}$-linear combination of $\vec{a}_1$, $\mathrm{D}\vec{a}_1$ and $\vec{a}_2$ by (5.4). More explicitly, $(\vec{a}_1, \vec{a}_2, \vec{a}_3)$ is a lattice vector of $T_A \oplus T_A \oplus T_A$, $A = J(C)$, belonging to the $\bar{\mathbb{Q}}$-linear subspace

$$W : \operatorname{diag}(\bar{\rho}u^2 + \rho v, -\bar{\rho}v, -\bar{\rho}v) \cdot \vec{x} - \bar{\rho}u \cdot \vec{y} - \vec{z} = 0 . \tag{5.12}$$

Without loss of generality we can assume that $u \neq 0$. Namely, the change of a typical basis $*\vec{\alpha}$ corresponds to a $\Gamma(\sqrt{-3})$-transform of the corresponding ball point, and $\Gamma(\sqrt{-3})$ does not fix any line through $\mathbb{B}$. Together with Lemma 5.10 we see that

$$u \in K \backslash 0. \tag{5.13}$$

The Corollary 5.7 of WUESTHOLZ's theorem yields a connected algebraic subgroup $G \neq 0$ of $A \times A \times A$ defined over $\bar{\mathbb{Q}}$ inside of $\exp(W)$. From (5.12) it follows immediately that

$$\dim G \leq \dim W = 6. \tag{5.14}$$

5.12. *G is a non-trivial subgroup of $A \times A \times A$.*

This means that G is not a product of some factors of $A \times A \times A$. For example we see that $W \cap (T_A \oplus T_A \oplus 0) < W$. Since $T_G \subseteq W$ it follows that $T_G \neq T_A \oplus T_A \oplus 0$, hence $G \neq A \times A \times 0$. The other cases are similar.

We denote by p_{ij} the projection $A \times A \times A \to A \times A$ forgetting the k-th component, $\{i, j, k\} = \{1, 2, 3\}$.

5.13. *$p_{12}(G)$ is a non-trivial algebraic subgroup of $A \times A$ (defined over $\bar{\mathbb{Q}}$).*

$p_{12}(G) \neq A \times A$ because $T_G \subseteq W$ and $W_{12} = \operatorname{Lie}(p_{12})(W) \not\subseteq T_A \oplus T_A$. Notice that

$$W_{12} : \operatorname{diag}(\bar{\rho}u^2 + \rho v, -\bar{\rho}v, -\bar{\rho}v) \cdot \vec{x} = \bar{\rho}u \cdot \vec{y} \tag{5.15}$$

by (5.12). Also $W_{12} \not\subseteq T_A \oplus 0$, $0 \oplus T_A$, hence $p_{12}(G) \neq A \times 0, 0 \times A$. Assume that $p_{12}(G) = 0$. Then $G \subseteq 0 \times 0 \times A$, hence $G = 0 \times 0 \times A$ because A is simple and $\dim G > 0$. This contradicts to 5.12.

We apply 5.2 and 5.3 to $p_{12}(G) \subset A \times A$. We get an isogeny φ in $\operatorname{End} A$ with $\operatorname{graph}(\varphi) \subseteq \operatorname{Lie}(p_{12}(G)) \subseteq W_{12}$. Since $\mathbb{Q} \otimes \operatorname{End} A = K$, there exists $k \in K \setminus 0$ such that $\varphi = \iota(k)$. Comparing (5.11) with (5.15) it is now easy to see that $\bar{\rho}u^2 + \rho v \in K$, hence $v \in K$ by (5.13). The Corollary 5.11 and Theorem 5.1 are proved. □

Remark 5.14. In some sense the ball coordinate v has the same right as u in the implication of Lemma 5.10. In fact, we can deduce more elementary and immediately Corollary 5.11 from Lemma 5.10 without using WUESTHOLZ's theorem again in the following manner:

Let C_τ be the PICARD curve in Lemma 5.10 we started with. The assumptions of the Lemma are preserved, if we change over from C_τ to $C_{\tau'}$, $\tau' \in \Gamma(\sqrt{-3})\tau$ because of the SCHOTTKY-TORELLI diagram (3.24). The point $\tau' \in \mathbb{B}$ has affine ball coordinates (u', v') or projective coordinates $(1 : u' : v') \in \mathbb{P}^2(\bar{\mathbb{Q}})$, respectively. From Lemma 5.10 we know that $u, u' \in K$. Assume that $v \notin K$. We have the freedom to choose $\gamma \in \Gamma(\sqrt{-3})$, $\tau' = \tau\gamma$, in a suitable manner in order to get a contradiction.

For this purpose we set

$$\gamma = \begin{pmatrix} a_1 & a_2 & a_3 \\ b_1 & b_2 & b_3 \\ c_1 & c_2 & c_3 \end{pmatrix} .$$

Part of the relation $(1, u, v)\gamma = \lambda(1, u', v')$ is the linear system of equations for v, λ

$$c_1 v - \lambda = -a_1 - b_1 u$$
$$c_2 v - u'\lambda = -a_2 - b_2 ucr$$

with coefficients in K. It is easy to find $\gamma \in \Gamma(\sqrt{-3})$ such that $u' \neq c_2/c_1$. Then our system has a unique solution, hence $v \in K$, which was to be proved.

Lemma 5.15. *If* $\tau \in \mathbb{B}(K)$, *then the Jacobian threefold* $J(C_\tau)$ *of the* PICARD *curve* C_τ *is not simple.*

Proof: We set $\tau = \mathbb{P}\mathfrak{a}$, $\mathfrak{a} \in K^3$, $\langle \mathfrak{a}, \mathfrak{a} \rangle < 0$ with the hermitian form $\langle , \rangle$ introduced in section 2.4, see (2.44). The orthogonal complement $\mathfrak{a}^\perp$ of $\mathfrak{a}$ in $\mathbb{C}^3$ is defined over K. This follows from the explicit description of $\langle , rangle$ in (2.47). Take a general unitary automorphism β of $\mathfrak{a}^\perp$ with coefficients in K (with respect to a K-basis of $\mathfrak{a}^\perp$) and let α be the linear endomorphism of $\mathbb{C}\mathfrak{a}$ extending $\mathfrak{a} \mapsto a\mathfrak{a}$, $a \in K\backslash 0$. The orthogonal sum $g = \alpha \textcircled{\perp} \beta$ belongs to $\mathbb{U}((2,1), K)$ and has $\tau \in \mathbb{B}$ as isolated fixed point. Then τ is a singular module by Definition 4.37 and $J(C_\tau)$ has decomposed complex multiplication by FEUSTEL's Proposition 4.39.

From the equivalence result 4.39 we used only the weak direction already proved in [34] by the author. In more detail, we found in the proof of Lemma 2.27 a symplectic embedding $\gamma \mapsto G$. It can be extended to $\mathbb{U}((2,1), K)$ in the same manner, explicitly by solving the equation (2.50) in G for given g. The image G of g lies in the group of symplectic similitudes

$$\mathbb{G}p(6, \mathbb{Q}) = \{H \in \mathbb{G}l_6(\mathbb{Q});\quad {}^tHIH = qI \quad \text{for a suitable} \quad q \in \mathbb{Q}\}$$

(compare with (2.22)). Our special g fixes $\tau = \mathbb{P}\mathfrak{a}$ and defines a non-trivial element in $\mathbb{Q} \otimes \operatorname{End} J(C_\tau)$. We have enough freedom to choose g such that G does not belong to $\iota(K)$ in $\mathbb{Q} \otimes \operatorname{End} J(C_\tau)$. So the endomorphism algebra of $J(C_\tau)$ cannot be isomorphic to K.

Now look back at Lemma 5.9. If $J(C_\tau)$ were simple, then $\mathbb{Q} \otimes \operatorname{End} J(C_\tau)$ would have to be a cubic field extension of K because we just excluded the other possible case. But g satisfies a K-reducible cubic equation over K coming from the characteristic polynomial of $g = \alpha \textcircled{\perp} \beta$. It is not difficult to see that this equation (over $\iota(K)$) holds also for $G \in \mathbb{Q} \otimes \operatorname{End} J(C_\tau)$. Therefore this endomorphism algebra cannot be a cubic extension of K, hence $J(C_\tau)$ is not simple. □

5.2 Transcendence at Non-Singular Non-Simple Algebraic Moduli

We want to extend Theorem 5.1 to the general case leaving the restriction for τ to be of simple type. As we will see, already the use of classical results is sufficient for the case that J_τ splits up to isogeny into the product of elliptic curves. But it could happen also that $\tau \in \mathbb{B}$ is a *module of mixed type.* This means, by definition, that $J_\tau \approx E \times B$, E an elliptic curve and B a simple abelian surface. Then we will delegate the result we are looking for to the following theorem due to SHIGA. In the mean time it has a natural generalization to the higher dimensions, see the end of this section. But we only need and prove the following special case:

Theorem 5.16 (SHIGA [75]). *If B is a simple abelian surface defined over $\bar{\mathbb{Q}}$ and E a polarization such that one, hence all (see below), corresponding period points lay in $\mathbb{H}_2(\bar{\mathbb{Q}})$, then B has complex multiplication.*

All period points of a principally polarized abelian variety A of dimension g fill a $Sp(2g, \mathbb{Z})$-orbit on $\mathbb{H}_g$. We denote it by $\omega(A, E)$, E the polarization. So we have

$$\omega(A, E) \subseteq \mathbb{H}_g(\bar{\mathbb{Q}}) \quad \text{or} \quad \omega(A, E) \cap \mathbb{H}_g(\bar{\mathbb{Q}}) = \emptyset \tag{5.16}$$

in any case. The same is true also for other (non-principal) polarizations. One has only to work with "modified symplectic groups". They are defined as in (2.22) but with Δ instead of I, where Δ is defined as I in (2.21) but with a diagonal matrix

$$D = \operatorname{diag}(d_1, \ldots, d_g), \quad d_i \in \mathbb{N}_+, \ d_1 \mid d_2 \mid \ldots \mid d_g$$

instead of E_g.

5.17. *The property $\omega(A, E) \subseteq \mathbb{H}_g(\bar{\mathbb{Q}})$ depends only on the isogeny class of (A, E).*

Proof: Let A and B be two isogeneous polarized abelian varieties. Since our property depends only on the isomorphy classes we can work with any representants. We choose period matrices Π, Σ representing the polarizations of A or B, respectively, such that the columns of $\Sigma = (\Sigma_1 \mid \Sigma_2)$ are natural multiples of the columns of $\Pi = (\Pi_1 \mid \Pi_2)$. Then it is clear that the period point $\Omega = \Pi_1^{-1}\Pi_2$ belongs to $\mathbb{H}_g(\bar{\mathbb{Q}})$ if and only if $\Sigma_1^{-1}\Sigma_2$ does. □

5.18. *If A is an abelian variety over $\bar{\mathbb{Q}}$, then its isogeny decomposition in simple factors, especially each factor itself, has a $\bar{\mathbb{Q}}$-model.*

Proof: We take first an isogeny decomposition

$$A \approx B_1 \times B_2 \times \ldots \times B_r \tag{5.17}$$

in simple factors B_i over $\bar{\mathbb{Q}}$. The endomorphism algebra $\mathbb{Q} \otimes \operatorname{End} B_i$ is a division algebra D_i. This is a precise criterion for simplicity. Also D_i is an invariant of the isogeny class of B_i. So B_i is not only simple over $\bar{\mathbb{Q}}$ but also over $\mathbb{C}$. So (5.17) is not only the unique (up to isogeny) decomposition in simple factors over $\bar{\mathbb{Q}}$ but also over $\mathbb{C}$. Moreover, "to be defined over $\bar{\mathbb{Q}}$" is an invariant property of isogeny classes. Therefore the simple factors of any isogeny decomposition (5.17) of A can be defined over $\bar{\mathbb{Q}}$. □

Take $B = B_1 \times \ldots \times B_r = \mathbb{C}^g/\Lambda_\Pi$ in the isogeny class of (A, E) with decomposition (5.17). Then $\Lambda_\Pi = \Lambda_{\Pi_1} \oplus \ldots \oplus \Lambda_{\Pi_r}$ and $B_i = \mathbb{C}^{g_i}/\Lambda_{\Pi_i}$. The period matrix Π consists of a diagonal block and a g-shifted diagonal block coming from the partitions of the Π_i into two squares. Writing $\Pi = (\Pi' \mid \Pi'')$ we see now that the period matrix $\Omega = \Pi'^{-1}\Pi''$ is composed by r diagonal blocks Ω_i of size g_i.

5.19. *The period point Ω belongs to $\mathbb{H}_g(\bar{\mathbb{Q}})$ if all Ω_i belong to $\mathbb{H}_{g_i}(\bar{\mathbb{Q}})$, respectively.*

Starting with a polarized abelian variety (A, E) with decomposition (5.17) we endowed the factors in a natural manner with polarizations such that 5.19 holds. We proved

Proposition 5.20. *Each polarized abelian variety (A, E) defined over $\bar{\mathbb{Q}}$ splits up to isogeny into a product of simple polarized abelian varieties $(B_1, E_1), \ldots, (B_r, E_r)$ defined over $\bar{\mathbb{Q}}$. Moreover, it holds that*

$$\omega(A, E) \subseteq \mathbb{H}_g(\bar{\mathbb{Q}}) \quad \text{iff} \quad \omega(B_i, E_i) \subseteq \mathbb{H}_{g_i}(\bar{\mathbb{Q}}), \quad i = 1, \ldots, r,$$

where $g = \dim A$, $g_i = \dim B_i$. □

For polarized abelian varieties (A, E) we want to compare the following three properties:

$(\bar{\mathbb{Q}})$ $\quad (A, E)$ is defined over $\bar{\mathbb{Q}}$;

$(\mathbb{H}(\bar{\mathbb{Q}}))$ $\quad \omega(A, E) \subseteq \mathbb{H}_g(\bar{\mathbb{Q}})$, where $g = \dim A$;

(DCM) $\quad A$ has decomposed complex multiplication.

Corollary 5.21. *For all polarized abelian varieties (A, E) over $\mathbb{C}$ (of dimension g) the implication*

$$(\bar{\mathbb{Q}}), \quad (\mathbb{H}(\bar{\mathbb{Q}})) \Longrightarrow (DCM) \tag{5.18}$$

is true if and only if it holds for all simple polarized complex abelian varieties (of dimension $\leq g$).

Proof: One direction is trivial. The other follows immediately from Proposition 5.20. Namely, under the assumptions of 5.20 and the implication (5.18) applied to the simple factors we see that each B_i has complex multiplication. But this means, by definition, that A has decomposed complex multiplication (DCM). □

Example 5.22. The implication (5.18) holds for elliptic curves. Indeed, assume that E is an elliptic curve defined over $\bar{\mathbb{Q}}$ with period point (module) $\tau \in \mathbb{H}(\bar{\mathbb{Q}})$ and E has no complex multiplication. The Theorem of SCHNEIDER (see Theorem 5.36 below) says that $j(\tau)$ is transcendental. This means that the moduli point of E in the moduli space $\mathbb{P}^1/S_3$ of elliptic curves is transcendental. But then also its preimage $(e_0 : e_1 : e_2) \in \mathbb{P}^1 = \mathbb{P}^2_0$ is. Therefore there is no WEIERSTRASS normal form

$$y^2 = 4(X - e_0)(X - e_1)(X - e_2)$$

of E with algebraic coefficients. Thus E cannot be defined over $\bar{\mathbb{Q}}$. □

Example 5.23. Let A be a polarized abelian variety of *elliptic decomposition type*, this means that the factors B_i of the decomposition (5.17) are elliptic curves. Assume that $\omega(A, E) \subseteq \mathbb{H}_g(\bar{\mathbb{Q}})$ and that A is defined over $\bar{\mathbb{Q}}$. The simple elliptic factors inherit both properties $(\bar{\mathbb{Q}})$ and $(\mathbb{H}(\bar{\mathbb{Q}}))$ by Proposition 5.17. From 5.22 it follows that all elliptic curves B_i have complex multiplication. This means that A has decomposed complex multiplication. Thus the implication (5.18) is true for all abelian varieties of elliptic decomposition type. □

We are able now to extend Theorem 5.1 to *(non-singular) modules* $\tau \in \mathbb{B}$ *of elliptic decomposition type*. This means that $J_\tau = J(C_\tau)$ is of elliptic decomposition type.

Corollary 5.24. *Let $\tau \in \mathbb{B}(\bar{\mathbb{Q}})$ be a non-singular module of elliptic decomposition type. Then $th(\tau) \in \mathbb{P}^2$ is transcendental.*

Proof: "Non-singular" means that J_τ is not of DCM-type, see Proposition 4.39. By assumption, the affine ball coordinates u, v of τ are algebraic. By (5.6) the period matrix $\Omega(u, v)$ of J_τ is also algebraic. In contraposition to (5.18) the implication

$$(\mathbb{H}(\bar{\mathbb{Q}})), \quad \text{not} \quad (DCM) \Longrightarrow \quad \text{not} \quad (\bar{\mathbb{Q}}) \qquad (5.19)$$

is correct for J_τ because of our assumptions and Example 5.23.

The point $th(\tau)$ is the moduli point of the PICARD curve

$$C_\tau : Y^3 = \Pi_{i=1}^4 (X - th_i(\tau))$$

by Theorem 3.33. From the implication (5.19) we know that J_τ has no $\bar{\mathbb{Q}}$-model. But then also C_τ has no $\bar{\mathbb{Q}}$-model, especially the curves isomorphic to C_τ are not defined over $\bar{\mathbb{Q}}$. Therefore $th(\tau) = (th_1(\tau) : th_2(\tau) : th_3(\tau) : th_4(\tau))$ is transcendental. □

Lemma 5.25. *The period points of a simple polarized abelian surface A cannot belong to $\mathbb{H}_2(K)$ for any quadratic number field K.*

Proof: The period point Ω comes from a period matrix Π of size 2×4. We have the relation

$$\Pi = (\mathfrak{a}_1, \mathfrak{a}_2 \mid \mathfrak{a}_3, \mathfrak{a}_4) = (\Pi_1 \mid \Pi_2) = \Pi_1(E_2 \mid \Omega)\,. \tag{5.20}$$

Now we suppose that Ω belongs to $\mathbb{H}_2(K)$ for a quadratic number field K. We construct a non-degenerate morphism from an elliptic curve into A. Then we get a contradiction to the assumption that A is simple.

Since Im $\Omega > 0$ we can write

$$\Omega = \begin{pmatrix} \lambda_1 & \lambda_2 \\ \lambda_2 & \lambda_3 \end{pmatrix}, \tag{5.21}$$

$$\lambda_2 = a + b\lambda,\ \ \lambda_3 = c + d\lambda,\ \ a, b, c, d \in \mathbb{Q},\ \ \lambda = \lambda_1 \in K\backslash\mathbb{Q}.$$

From the above relation (5.20) we get

$$\lambda\mathfrak{a}_1 + b\lambda\mathfrak{a}_2 \in \mathbb{Q}\Lambda,$$

where Λ is the $\mathbb{Z}$-lattice in $\mathbb{C}^2$ generated by the columns $\mathfrak{a}_i$ of Π, $i = 1, \ldots, 4$, $A \cong \mathbb{C}^2/\Lambda$. Observe that

$$\mathfrak{a} = \mathfrak{a}_1 + b\mathfrak{a}_2 \neq 0$$

because the columns of Π are $\mathbb{Q}$-linearly independent. Multiplying, if necessary, λ and $\mathfrak{a}$ with suitable natural numbers we find $\mu \in \mathcal{O}_K$ and $\mathfrak{b} \in \Lambda\backslash 0$ such that $\mu\mathfrak{b} \in \Lambda$. With $\mathcal{O} = \mathbb{Z}[\mu]$ we get $\mathcal{O}\mathfrak{b} \subset \Lambda$. The commutative diagram with exact rows

$$\begin{array}{ccccccccc} 0 & \longrightarrow & \mathcal{O}\mathfrak{b} & \longrightarrow & \mathbb{C}\mathfrak{b} & \longrightarrow & \mathbb{C}/\mathcal{O}\mathfrak{b} & \longrightarrow & 0 \\ & & \downarrow & & \downarrow & & \downarrow & & \\ 0 & \longrightarrow & \Lambda & \longrightarrow & \mathbb{C}^2 & \longrightarrow & \mathbb{C}^2/\Lambda & \longrightarrow & 0 \end{array}$$

yields a splitting of $A \cong \mathbb{C}^2/\Lambda$ as were looking for. □

5.26. The proof shows that the elements of the first row of the period matrix Ω cannot belong simultaneously to a quadratic number field K under the assumptions of the lemma.

Proposition 5.27 ([75]). *The implication* (5.18) *is true for polarized abelian surfaces A.*

Proof: If A is not simple, then it is of elliptic decomposition type. We are therefore in the situation of Example 5.23.

Now assume that A is simple and has the properties $(\bar{\mathbb{Q}})$,$(\mathbb{H}(\bar{\mathbb{Q}}))$. With the notations of (5.20) we get the vector relation

$$\lambda_1 \mathfrak{a}_1 + \lambda_2 \mathfrak{a}_2 - \mathfrak{a}_3 = 0 \tag{5.22}$$

with coefficients λ_i in $\bar{\mathbb{Q}}$ by assumption $(\mathbb{H}(\bar{\mathbb{Q}}))$. The abelian variety $B = A \times A \times A$ is defined over $\bar{\mathbb{Q}}$ by assumption $(\bar{\mathbb{Q}})$. Consider the $\bar{\mathbb{Q}}$-linear subspace

$$W : \lambda_1 \vec{z} + \lambda_2 \vec{w} - \vec{u} = 0 \tag{5.23}$$

in $T_B = T_A \oplus T_A \oplus T_A$, where $\vec{z}, \vec{w}, \vec{u}$ denote vector variables of the summands T_A of T_B. The lattice point $(\mathfrak{a}_1, \mathfrak{a}_2, \mathfrak{a}_3) \in T_B = \mathbb{C}^2 \times \mathbb{C}^2 \times \mathbb{C}^2$ belongs to W. By the Corollary 5.7 of WUESTHOLZ's Theorem 5.5 there exists a non-finite algebraic subgroup H of $B = A \times A \times A$ with the properties described in diagram (5.2) but with B instead of A there. H is defined over $\bar{\mathbb{Q}}$.

Denote by H_{ij} or W_{ij}, $1 \leq i < j \leq 3$, the projections of H or W onto the ij-th factor $A \times A$ of B or onto the summand $T_A \oplus T_A$ of T_B, respectively. Restricting diagram (5.2) (with B instead of A) we get commutative diagrams

$$\begin{array}{ccc} T_{A_{ij}} & \xrightarrow{\exp} & A_{ij} = A \times A \\ \uparrow & & \cup\!\shortmid \\ | & & \exp(W_{ij}) \\ | & & \cup\!\shortmid \\ w_{ij} \in T_{H_{ij}} & \longrightarrow & H_{ij} \end{array} \tag{5.24}$$

w_{ij} the projection of $w = (\mathfrak{a}_1,\ \mathfrak{a}_2,\ \mathfrak{a}_3)$. In the diagram we can substitute w_{ij} by qw_{ij} for each $q \in \mathbb{Q}$. So H_{ij} cannot be finite, dim $H_{ij} > 0$. On the other hand we get from (5.23) the equations for

$$\mathrm{W}_{13} : \vec{\mathrm{u}} = \lambda_1 \vec{\mathrm{z}} \quad , \qquad \mathrm{W}_{23} : \vec{\mathrm{u}} = \lambda_2 \vec{\mathrm{w}} \ . \tag{5.25}$$

Since A is simple the period point Ω cannot be a diagonal matrix, hence $\lambda_2 \neq 0$; $\lambda_1 \neq 0$ follows from Im $\Omega > 0$. The equations (5.25) show that $\dim W_{13} = \dim W_{23} = 2$, hence $\dim H_{13}$, $\dim H_{23} \leq 2$, by diagram (5.24). Together with (5.2) we get

$$\dim \mathrm{H_{ij}} = 2 \quad , \quad \mathrm{T_{ij}} = \mathrm{W_{ij}} \quad \text{for} \quad (\mathrm{i,j}) = (1,3), (2,3). \tag{5.26}$$

Especially, our subgroups H_{ij} of $A \times A$ are not trivial in the sense of 5.4. By 5.3 they define elements φ_{13}, $\varphi_{23} \in \mathbb{Q} \otimes \mathrm{End}\, A$. Their representations in $\mathbb{C}^2 = T_A$ are given by

$$\mathfrak{a} \longmapsto \lambda_1 \mathfrak{a} \quad , \quad \mathfrak{a} \longmapsto \lambda_2 \mathfrak{a}, \tag{5.27}$$

respectively. This follows from (5.25) and (5.26). Moreover, φ_{13} and φ_{23} commute with each other because their diagonal representations (5.27) do.

The endomorphism algebra of A is a division algebra because A is simple. It contains the subfield $\mathbb{Q}(\varphi_{13}, \varphi_{23}) \cong \mathbb{Q}(\lambda_1, \lambda_2)$. Since λ_1 is not real and each subfield of $\mathbb{Q} \otimes \operatorname{End} A$ divides $2 \cdot \dim A = 4$ we get the relations of division

$$2 \quad | \quad [\mathbb{Q}(\lambda_1) : \mathbb{Q}] \quad | \quad [\mathbb{Q}(\lambda_1, \lambda_2) : \mathbb{Q}] \quad | \quad 4.$$

If $[\mathbb{Q}(\lambda_1, \lambda_2) : \mathbb{Q}] = 4$, then A has complex multiplication by definition. If not, then λ_2 belongs to the quadratic number field $\mathbb{Q}(\lambda_1)$. But then A is not simple by Lemma 5.25 and Remark 5.26 in contradiction to our assumptions. □

Now we can repeat our inductive argument in Example 5.23 in order to prove

Cororllary 5.28. *For complex polarized abelian threefolds of mixed or elliptic decomposition type the implication* (5.18) *is true.* □

We can repeat also the argument in the proof of Corollary 5.24 and obtain

Corollary 5.29. *Let $\tau \in \mathbb{B}(\bar{\mathbb{Q}})$ be a non-singular module of mixed decomposition type. Then $th(\tau) \in \mathbb{P}^2$ is transcendental.* □

Now we put together the results of Theorem 5.1 and of the Corollaries 5.24, 5.29 to get the final common result.

Theorem 5.30. *Let $\tau \in \mathbb{B}(\bar{\mathbb{Q}})$ be a non-singular module. Then $th(\tau) \in \mathbb{P}^2$ is a transcendental point.* □

At the end we remark that SHIGA [77] announced he was able to extend the Proposition 5.9 to all dimensions. Without proof we quote

Theorem 5.31 (SHIGA [77]). *The implication* (5.18) *is true for all complex polarized abelian varieties.*

5.3 Some More History

After the celebrated transcendence proofs for e and π by HERMITE [29] or LINDEMANN [50], respectively, HILBERT invited the mathematicians in his seventh problem, entitled "Irrationality and Transcendence of Certain Numbers", to work on further progress. The basic problem to prove the transcendence of a^β for algebraic numbers a, β, $a \neq 0, 1$, β irrational, has been solved independently by SCHNEIDER [65] and GELFOND [24] in 1934. The transcendence of the function $e^{i\pi z}$ at irrational algebraic values follows immediately using $e^\pi = i^{-2i}$.

Implicitly HILBERT expected more: "That certain special transcendental functions, important in analysis, take algebraic values for certain algebraic arguments, seems to us particularly remarkable and worthy of thorough investigation. "..." It is certain that the solution of these and similar problems must lead us to entirely new insight into the nature of special irrational and transcendental numbers."

We would like to formulate a general *homogeneous transcendence problem* for transcendental holomorphic maps in higher dimension generalizing the seventh HILBERT problem in close (complementary) connection with HILBERT's twelfth problem. For this purpose we consider (open analytic) submanifolds or subvarieties of the complex projective spaces $\mathbb{P}^N = \mathbb{P}^N(\mathbb{C})$. Let $V \subseteq \mathbb{P}^m$, $W \subseteq \mathbb{P}^n$ be two of them. Remember that a holomorphic map

$$f = (f_0 : f_1 : \ldots : f_n) : V \longrightarrow W$$

is called *transcendental* if it is not the restriction of an algebraic (rational) map. For a subfield K of $\mathbb{C}$ the K-points of $\mathbb{P}^N, V, W, \ldots$ are defined and denoted by

$$\mathbb{P}^N(K) = \{(x_0 : x_1 : \ldots : x_N) \in \mathbb{P}^N(\mathbb{C}); x_i \in K, \; i = 0, \ldots, N\},$$
$$V(K) = V \cap \mathbb{P}^m(K), \ldots$$

The field of algebraic numbers is denoted by $\bar{\mathbb{Q}}$ as usual. The $\bar{\mathbb{Q}}$-points of $\mathbb{P}^N, V, \ldots$ are called *algebraic* or *arithmetic* points of $\mathbb{P}^N$, V, ..., respectively. The non-algebraic points are called *transcendental.*

Problem 5.32. Find explicitly transcendental (projective holomorphic) maps $f = (f_0 : \ldots : f_n) : V \longrightarrow W$, where you can characterize

$$(a) \qquad T(f) = \{P \in V(\bar{\mathbb{Q}}); f(P) \text{ transcendental}\}$$

or/and

$$(b) \qquad T(f^{-1}) = \{Q \in W(\bar{\mathbb{Q}}); f^{-1}(Q) \cap V(\bar{\mathbb{Q}}) = \emptyset\}$$

by simple algebraic/arithmetic properties.

Observe that there is already a fractional statement in HILBERT's 7-th problem itself: "If, in an isosceles triangle, the ratio of the base angle to the angle at the vertex be algebraic but not rational, the ratio between base and side is always transcendental."

Some more historical development lead us to our two-dimensional model. SIEGEL [81] started in 1932 to study transcendental properties of elliptic integrals. He proved that among the periods of an elliptic integral of first kind with algebraic coefficients at least one is transcendental. Three years later SCHNEIDER [66] proved that the non-vanishing periods are transcendental as well for elliptic integrals of first as of second kind (with algebraic coefficients). Moreover, from transcendental studies of values of the WEIERSTRASS $\wp$-function it follows that the elliptic modular function $j(\tau)$ has transcendental values at algebraic arguments which are not imaginary quadratic.

Proofs of the above results can be found in SCHNEIDER's book [68], see also [82]. At the end of the book SCHNEIDER placed 8 problems. We recall two of them.

5.33 SCHNEIDER's **second problem.** *The theorem on transcendence of values of the modular function $j(\tau)$ should be proved by direct investigations of the transcendental modular function not using the results on $\wp$-functions.*

5.34 SCHNEIDER's **fourth problem.** *The transcendence results on elliptic integrals of first and second kind should be generalized as far as possible to analogous results on abelian integrals.*

In a private communication WUESTHOLZ explained me the reduction of the transcendence proof for the elliptic modular function $j(\tau)$ to his subgroup Theorem 5.5. Unfortunately, the subgroup theorem needs the exponential map from the tangent spaces to elliptic curves realized by WEIERSTRASS $\wp$-functions explicitly. In this way SCHNEIDER's second problem is touched but not solved. It has been also published in SHIGA's paper [75], and we reproduce the short proof at the end of this section. An important idea to intervene consequently algebraic group theory goes back to S. LANG [44].

A general transcendent result for abelian integrals of algebraic type has been established recently by WUESTHOLZ, see the last proposition in Section 1 of [90]: The non-vanishing period integrals of first kind are transcendental. But for the ratios there exists the obstruction coming from complex multiplications.

We call an *abelian integral* $\int \omega$ *algebraic*, if it is taken on an algebraic curve C defined over $\bar{\mathbb{Q}}$ and ω is an algebraic differential form on C. Restricting to PICARD integrals (3.23) we found the following two-dimensional contribution supporting the second part of the general problem 5.32 by special solution:

Theorem 5.35. *Let $\mathcal{O}$ be the ring of* EISENSTEIN *integers. For three algebraic* PICARD *integrals I_i on the same smooth* PICARD *curve C along $\mathcal{O}$-linearly independent cycles the $\mathbb{P}^2$-point $(I_0 : I_1 : I_2)$ is transcendental if and only if the curve is not of DCM-type.*

Proof: Obviously, we can restrict ourselves to a typical $\mathcal{O}$-basis of $H_0(C, \mathbb{Z})$. Then $\tau = (I_0 : I_1 : I_2)$ lies in the ball $\mathbb{B}$ and $\mathrm{th}(\tau) \in \mathbb{P}^2$ is the moduli point of C by Theorem 3.31 (effective TORELLI) and the presentation of (3.26) by *th*, see Theorem 3.35. By assumption C is algebraically defined, this means that $\mathrm{th}(\tau) \in \mathbb{P}^2(\bar{\mathbb{Q}})$. The curve $C \cong C_\tau$ is not of DCM-type if τ is a non-singular module, by definition. Assume that C is not of *DCM*-type. Then from Theorem 5.30 it follows that $\tau \notin \mathbb{B}(\bar{\mathbb{Q}})$, that means $\tau = (I_0 : I_1 : I_2)$ is transcendental. If C is DCM, then τ is an isolated fixed point of an element $\gamma \in \mathbb{U}((2,1), K)$, see (4.39). Therefore $\tau \in \mathbb{B}(\bar{\mathbb{Q}})$ in this case. □

Conversely, $\mathrm{th}(\tau)$ is algebraic for singular moduli $\tau \in \mathbb{B}$, see Chapter 4, but transcendental by Theorem 5.30, if $\tau \in \mathbb{B}(\bar{\mathbb{Q}})$ is non-singular. So $f = \mathrm{th}$ is also a special solution of 5.32 (a).

There is some actual work on E-functions (see for example BEUKERS [6], WOLFART [7], SHIDLOVSKY [72]) introduced and first investigated by SIEGEL [80].

The most important come out as solutions of a system of ordinary homogeneous linear differential equations with rational functions as coefficients.

As we know (see Theorem 3.23 and (3.23)), the PICARD integrals can be considered as (multivalued) analytic functions on the space $\mathbb{P}^2\backslash\bigstar\!\!\!\!\bigtriangleup$ solving the special Fuchsian system (3.10) of linear partial differential equations of two variables. So it seemed to us interesting enough to present our comprehensive intrinsic transcendence proof concentrated on this example.

For the sake of self-contents as much as possible we present the direct proof of the following theorem.

Theorem 5.36 (SCHNEIDER [66], [68]). *The elliptic modular function $j(\tau)$ has transcendental values at non-quadratic algebraic arguments $\tau \in \mathbb{H}$.*

Proof: (WUESTHOLZ, private communication; SHIGA [75]). Assume that τ and $j(\tau)$ are algebraic. Then there is an elliptic curve (with modular invariant $j(\tau)$) $E \cong \mathbb{C}/\mathbb{Z}+\mathbb{Z}\tau$ defined by a plane cubic equation with algebraic coefficients. We are in the situation of Corollary 5.7 setting $A = E \times E$. Namely, $\exp(w)$ is a torsion point on A for $w = (w_1, w_2) \in T_A$, w_1, w_2 period integrals, $\tau = w_2/w_1$, hence $\exp(w) \in A(\bar{\mathbb{Q}})$. Obviously, w belongs to the one-dimensional linear $\bar{\mathbb{Q}}$-subspace W of T_A defined by $W : y = \tau x$. Applying (5.7) we obtain a diagram (5.2) with a one-dimensional non-trivial $\bar{\mathbb{Q}}$-defined algebraic subgroup H of $A = E \times E$. Now it follows immediately from (5.2), (5.3) and (5.4) that E has complex multiplication or, equivalently, τ is an imaginary quadratic algebraic number. This contradicts our assumption. □

At the end of this chapter we present a reformulation of the main results (5.30) and (5.35) in terms of points of SHIMURA varieties. We refer back to section 4.9 and recall to the basic notions there:

$$\mathbb{S} = R_{\mathbb{C}/\mathbb{R}}\mathbb{G}_m, \quad \mathbb{G}/\mathbb{Q} : \ \mathbb{G}(\mathbb{Q}) = \mathbb{GU}((2,1), K),$$

$$\mathbb{B} \cong \{h = g^{-1}h_0 g; \ \ g \in \mathbb{G}(\mathbb{R})\},$$

where "points" h are defined as (non-trivial) group homomorphisms

$$h = u_{\mathbb{R}} \circ h' : \quad \mathbb{S} \longrightarrow T_{\mathbb{R}} \hookrightarrow G_{\mathbb{R}},$$

T the image torus of h, see Chapter 4, (4.43). *Special points* are defined by the condition: u (hence T) is defined over $\mathbb{Q}$, see 4.56.

If u is defined over $\bar{\mathbb{Q}}$, then we call h an *algebraic point.* With the notations of Chapter 4, (4.41) we call a point of the canonical model ${}_{\mathbf{K}}Sh/K$ corresponding to

$$P = \mathbf{K}_\infty \times \mathbf{K}\backslash(h, \gamma)/\mathbb{G}(\mathbb{Q}) \in \mathbf{K}\backslash\mathbb{B} \times \mathbb{G}(\mathbb{A}^f)/\mathbb{G}(\mathbb{Q})$$

special or *universal* algebraic, if h is special or algebraic, respectively. Then for $K = \mathbb{Q}(\sqrt{-3})$ we have with obvious notations the following result:

Theorem 5.37 (Characterization of special points). *The point P of ${}_{\mathbf{K}}Sh(\mathbb{C})$ is special, if and only if it is algebraic and universal algebraic.*

Proof: Universal algebraic means that the "uniformizing point" h belongs to $\mathbb{B}(\bar{\mathbb{Q}})$. Since for special $\mathbf{K}$ we have ${}_{\mathbf{K}}Sh(\mathbb{C}) = \mathbb{P}^2(\mathbb{C})/S_4$. P is algebraic if and only if it comes from $\mathbb{P}^2(\bar{\mathbb{Q}})$. On the other hand special points on $\mathbb{P}^2$ are precisely the DCM-points. Now apply (5.30) and (5.35). □

Remark 5.38. Observe that a $\mathbb{Q}$-condition is equivalent to two $\bar{\mathbb{Q}}$-conditions.

In the meantime the transcendence result has been generalized to all SHIMURA varieties. We refer to P.B. COHEN, H. SHIGA, J. WOLFART, *Criteria for complex multiplication and transcendence properties of automorphic functions*, Math. Sem. der J.-W.-Goethe-Universität Frankfurt/M, Sept. 1993.

6 Arithmetic Surfaces of Kodaira-Picard Type and some Diophantine Equations

6.1 Introduction

We define curves and arithmetic surfaces of KODAIRA-PICARD type. Using ideas and results of VOJTA [89] we show that the Asymptotic FERMAT Theorem holds, if PARSHIN's analogue of the BOGOMOLOV-MIYAOKA-YAU inequality is true for arithmetic surfaces of KODAIRA-PICARD type. This gives a nice motivation for a finer study of the arithmetic and geometry of PICARD curves connected with arithmetic points of the PICARD modular surface of EISENSTEIN numbers and of the complex unit ball covering this surface. Until now a lot of preparatory work has already been done concerning PICARD modular forms in the framework of HILBERT's 12-th Problem.

The reader should be familiar with the basic notions of the theory of arithmetic surfaces [49]. For his convenience we list the most important definitions of heights and explicit formulas connected with modular forms.

One knows that the FALTINGS-NOETHER formula [21] and heights involve modular forms. We hope that the explicit knowledge of PICARD modular forms as theta constants and special values of them will interact fruitfully in the near future.

The final link will be established with the following version of the

6.1 Asymptotic FERMAT **Theorem (Conjecture)**[1]**:** *There is a positive constant C such that for all natural numbers $n > C$ the* FERMAT *equation*

$$X^n + Y^n = Z^n \tag{6.1}$$

has only trivial $\mathbb{Q}$-rational solutions.

6.2 Arithmetic Surfaces and Curves of KODAIRA-PICARD Type

Definitions 6.2. An *arithmetic surface of* PICARD *type* is an arithmetic surface X/Y, $Y = \operatorname{Spec} \mathcal{O}$, $\mathcal{O} = \mathcal{O}_K$ the ring of integers of number field K, such that general fibre X_K over K is a smooth PICARD curve, that means it is isomorphic to a smooth plane algebraic curve with affine equation

$$Y^3 = X^4 + G_2X^2 + G_3X + G_4, \quad G_i \in \mathbb{C}\,. \tag{6.2}$$

1) Perhaps it is now a theorem by the work of A. Wiles.

A (smooth compact complex algebraic) curve D is of KODAIRA-PICARD *type*, if it is either a biquadratic cover of a smooth PICARD curve C branched over exactly one point $P \in C$, a quadratic unramified cover of a PICARD curve or it is a PICARD curve itself. In these cases we call D also a KODAIRA-PICARD curve of *biquadratic, quadratic* or of *simple* (covering) type, respectively.

An arithmetic surface X/Y is of (biquadratic/quadratic/simple) KODAIRA-PICARD *type*, if its general fibre is.

Remark 6.3. PARSHIN transferred a construction of special complex algebraic surfaces due to KODAIRA to arithmetic surfaces in order to deduce the MORDELL conjecture from SHAFAREVIČ's conjecture (PARSHIN trick). The application to the diophantine equations

$$ZY^3 = X^4 + G_z X^2 Z^2 + G_3 X Z^3 + G_4 Z^4 \tag{6.3}$$

of PICARD type at a solution point yields an arithmetic surface of KODAIRA-PICARD type with general fibre C'' of the same biquadratic type. For details we refer to [89], where one finds the properties we need in the Definitions 6.2. Smooth PICARD curves have genus 3. From HURWITZ's genus formula it follows that a quadratic intermediate covering C'/C is unramified, C' has genus 5, C''/C' is branched above precisely two points, and C'' has genus 10. There should be a nice moduli space of curves of biquadratic KODAIRA-PICARD type (a finite quotient of a cubic cover of a simple rational threefold?). Via limit points of the open part corresponding to smooth curves one can extend the Definitions 6.2 to singular cases. For precise moduli interpretations similar to those we expect, one should pay attention to HUNT's paper [40].

Remark 6.4. As in the case of elliptic curves the constants G_i in (6.2) or (6.3) can be considered as special values of (PICARD) modular forms on the complex 2-ball $\mathbb{B}$. We refer to Chapter 3 and remember that the G_i are the S_4-elementary symmetric polynomials of simple linear combinations th_i of third power of theta constants on $\mathbb{H}_3$ defined in (3.41) restricted to the ball along the embedding $\star$ of the SCHOTTKY-TORELLI diagram (3.24).

6.3 Heights

6.5 WEIL **heights.** WEIL*'s height function on* $\mathbb{P}^n$ is the correspondence

$$\begin{aligned} h &: \mathbb{P}^n(\overline{\mathbb{Q}}) \longrightarrow \mathbb{R}, \\ h(x) &= \sum_{v \in M_K} [K_v : \mathbb{Q}_v] \cdot \log \max_j \{|x_i|_v\}, \end{aligned} \tag{6.4}$$

where $x = (x_0 : x_1 : \ldots : x_n) \in \mathbb{P}^n(K)$, $K \subset \overline{\mathbb{Q}}$ a number field (of finite degree), and M_K denotes the set of all (finite and infinite) valuations of K.

Let V/K be a projective variety, L a very ample line bundle on V and $\varphi_s : V \longrightarrow \mathbb{P}^n$ a projective embedding corresponding to a projective choice $s = (s_0 : s_1 : \ldots : s_n)$ of basic global sections L. Then one defines a WEIL *height function on* V by restriction:

$$h_s : V(K) \longrightarrow \mathbb{R},\ P \mapsto h(\varphi_s(P)). \tag{6.5}$$

A WEIL *height* on V is an equivalence class of WEIL height functions on V modulo the addition of restricted real functions on $V(\overline{\mathbb{Q}})$. For a very ample class $L \in \operatorname{Pic} V$ the WEIL height h_L is correctly defined by any representant h_s. The definition of WEIL heights h_L extends to all $L \in \operatorname{Piv} V$ via decompositions $L = L_1 \otimes L_2^{-1}, L_1, L_2$ very ample. Working with representants we note the most important properties (see e.g. [71]):

(add) $h_L = h_{L_1} + h_{L_2} + O(1)$ for $L \cong L_1 \otimes L_2$;

(fin) $\{P \in V(K); h_L(P) \leq C, [K : \mathbb{Q}] \geq d\}$
is finite for given $C, d > 0$ and ample L;

(func) $h_{f*L} = h_L \circ f + O(1)$
for $L \in \operatorname{Pic} V,\ f : W \longrightarrow V$ a morphism.

We write $f = g + O(1)$ if the difference $f - g$ of the two real functions f, g is restricted.

6.6 FALTINGS **degree.** Fix a number field K and set $R = \mathcal{O}_K$. A *line bundle on* $\operatorname{Spec} R$ is a projective R-module M of rank 1. We denote the set of archimedean metrics of K by M_K^∞. Let μ be a set of metrics $||\cdot||_v$ on $M \otimes K, v \in M_K^\infty$. The pair (M, μ) is called *metrized line bundle* and

$$\deg(M, \mu) = \log \sharp(M/Rm) - \sum_{v \in \mu} [K_v : \mathbb{Q}_v] \log ||m||_v, m \in M \backslash O, \tag{6.6}$$

the FALTINGS *degree* of (M, μ). It does not depend on the choice of m. If $\mu = M_K^\infty$ then we write $\deg M$.

Let V/K be a projective variety, $L \in \operatorname{Pic} V$. If L is endowed with v-adic metric $||\cdot||_v$ on $L \otimes K_v$ for each $v \in M_k^\infty$, then L is said to be a *metrized line bundle.* We assume that each norm $||\cdot||$ varies continuously on $V(K_v)$ w.r.t. sections over open subsets $U(K_v)$ of $V(K_v)$, see [83], p. 162. If $V = \mathcal{V} \otimes K$ and $L = \mathcal{L} \otimes K$ are general fibres of a model $\mathcal{V}/\operatorname{Spec} R$ or for $\mathcal{L} \in \operatorname{Pic}(\mathcal{V})$, respectively, then the metrization of L induces metrizations of pullbacks $\mathfrak{p}^*\mathcal{L}$ along sections $\mathfrak{p} \in \mathcal{V}(\operatorname{Spec} R)$. Since $\mathfrak{p}^*\mathcal{L}$ is a metrized line bundle, its FALTINGS degree is well-defined.

Proposition 6.7 (see SILVERMAN, [85], p. 271). *With the above notations and* $P = \mathfrak{p} \otimes K \in V(K)$ *it holds that*

$$h_L(P) = \deg \mathfrak{p}^*(\mathcal{L})/[K:\mathbb{Q}] + O(1). \tag{6.7}$$

Remark 6.8. *O(1) depends on the models* $\mathcal{V}, \mathcal{L}$ *but not on* $P \in V(K)$.

6.9 Modular height. Let A/K be an abelian variety of dimension g with NERON *model* $\mathcal{A}/\operatorname{Spec} R$ (see [47]), $\omega_{A/K} = \wedge^g \Omega^1_{A/K}$. For each $v \in M_K^\infty$ one defines a metric on $\omega_{A/K}$ by

$$||\alpha||_v = [(i/2)^g \int_{A_v(\mathbb{C})} \alpha \wedge \overline{\alpha}]$$

for any $\alpha \in H^0(A_v(\mathbb{C}), \omega_{A/K} \otimes \overline{K}_v), A_v = A \otimes K_v, \overline{K}_v = \mathbb{C}$. The zero point $0 \in A(K)$ extends to the zero section $\zeta \in \mathcal{A}(\text{spec R})$. The *modular height of* A/K is defined via FALTINGS degree by

$$h_{\text{mod}}(A/K) = \deg \zeta^* \omega_{\mathcal{A}/R}/[K:\mathbb{Q}]. \tag{6.8}$$

Remark 6.10. If one uses only semistable models $\mathcal{A}$ (possible via extension of definition field), then $h_{\text{mod}}(A/K)$ does not depend on the choice of K, and the notation $h_{\text{mod}}(A)$ is justified.

The name "modular height" is justified by a close relation with (WEIL) heights on the quasiprojective moduli space $\mathcal{A}_g/\mathbb{Q}$, $\mathcal{A}_g \subset \mathbb{P}^n$ say, of principally polarized complex abelian varieties (A, E) of dimension g. If A is defined over $\overline{\mathbb{Q}}$, then the moduli point of (A, E) is denoted by $j(A, E) \in \mathcal{A}_g(\overline{\mathbb{Q}})$.

Theorem 6.11 (FALTINGS, [22]). There are universal constants $c_i > 0$ such that for all principally polarized abelian varieties $(A, E)/K$ of dimension g with semistable reduction everywhere it holds that

$$c_1 h(j(A,E)) - c_2 < h_{\text{mod}}(A/K) < c_3 h(j(A,E)) + c_4 \,.$$

6.12 (SILVERMAN, [84], p. 254). *For elliptic curves* E/K, $E_v(\mathbb{C}) \cong \mathbb{C}/\mathbb{Z} + \mathbb{Z}\tau_v$, $v \in M_K^\infty$, *it holds that*

$$[K:\mathbb{Q}]h_{\text{mod}}(E/K) = \frac{1}{12}\left\{ \log|N_{K/Q}(\triangle_{E/K})| - \sum_{v \in M_K^\infty} \varepsilon_v \cdot \log \frac{|\triangle(\tau_v)|(Im\tau_v)^6}{(2\pi)^{12}} \right\}.$$

Here $\triangle_{E/K}$ denotes the minimal discriminant of E/K. This is an ideal of $\mathcal{O}_K$. The other $\triangle$ is the normalized modular cusp form

$$\triangle(\tau) = q_\tau \prod_{n=1}^{\infty} (1 - q_\tau^n)^{24}, \tau \in \mathbb{H} \text{ upper half plane,}$$

$$q_\tau = \exp(2\pi i \tau), \varepsilon_v = 1 \text{ if } K_v = \mathbb{R} \text{ or } 2 \text{ if } K_v = \mathbb{C}.$$

6.13 FALTINGS **height.** Let C/K be a smooth algebraic curve of genus $g, R = \mathcal{O}_K$, X/R an arithmetic surface model of C/K. We assume that X/R is semistable. ARAKELOV defined a model $\mathbf{J}/\operatorname{Spec} R$ of the Jacobian variety $\mathbf{J}(C)/K$ of C, and FALTINGS defined

$$h_{Fal}(X/R) = h_{\text{mod}}(\mathbf{J}/R). \tag{6.9}$$

Let $\pi : X \longrightarrow \operatorname{Spec} R$ be the projection, $\omega_{X/Y}$ the relative canonical bundle, $\pi_*\omega_{X/Y}$ its direct image on $\operatorname{Spec} R$. Working with admissible metrics (see [49]) one defines

$$\deg \pi_*\omega_{X/R} = \deg \det \pi_*\omega_{X/R}.$$

It turns out that

$$h_{Fal}(X/R) = \deg \pi_*\omega_{X/R}. \tag{6.10}$$

6.14 ARAKELOV **heights.** We fix $K, Y = \operatorname{Spec} R$ and an arithmetic surface X/Y with general fibre $X_K = C/K$. The group of ARAKELOV *divisors* of X/Y is denoted by $\operatorname{Div}_{Ar}(X/Y)$ and the corresponding class group by $\operatorname{Pic}_{Ar}(X/Y)$. On both groups the ARAKELOV *intersection product* $(\cdot)$ is defined, the latter induced by the former:

$$(\cdot) : \operatorname{Div}_{Ar}(X/Y) \times \operatorname{Div}_{Ar}(X/Y) \to \mathbb{R}$$

Each point $P \in C(\overline{\mathbb{Q}})$ extends to a horizontal (irreducible ARAKELOV) divisor $\mathbb{E}_P$ on X/Y supported by the ZARISKI closure of P in X. The definition field of P is denoted by $K(P)$. If $\mathbb{D}$ is an ARAKELOV divisor, then we denote by D its restriction to X_K, defined by the pullback of the line bundle $\mathcal{O}(\mathbb{D})$ belonging to $\mathbb{D}$. For the next result we refer to [89] again.

Proposition 6.15. *With above notations the correspondence*

$$h_{\mathbb{D}} : X_K(\overline{\mathbb{Q}}) \longrightarrow \mathbb{R},\ P \mapsto (\mathbb{D} \cdot \mathbb{E}_p)/[K(P) : \mathbb{Q}]$$

is a height function on C belonging to $D \in \operatorname{Div} C$:

$$h_{\mathbb{D}} = h_D + O(1).$$

Definition 6.16. The ARAKELOV *height (function)*

$$h_{Ar} = h_{Ar,X/Y} : X_K(\overline{\mathbb{Q}}) \longrightarrow \mathbb{R}$$

is defined by $h_{Ar} = h_{\omega_{X/Y}}$.

Corollary 6.17.

$$h_{Ar,X/Y} = h_{\omega_C} + O(1).$$

6.18 FALTINGS-NOETHER **formula.** We need some local invariants of an arithmetic surface X/Y. They have been introduced by the transfer of M. NOETHER's formula for algebraic surfaces to arithmetic surfaces due to FALTINGS. We have first to explain the arithmetic surface analogon of the arithmetic genus. For a finite R-module M metrized at infinity one defines

$$\chi(M,R) = \chi(M,\mathbb{Z}) - \operatorname{rank}_R(M) \cdot \chi(R,\mathbb{Z}),$$

where

$$\chi(M,\mathbb{Z}) = -\log \operatorname{Vol}(M_{\mathbb{R}}/M) + \log \sharp M_{tor}, \; M_{\mathbb{R}} = M \otimes \mathbb{R} = \prod_{v \in M_K^{\infty}} (M \otimes K_v).$$

The volume is taken with respect to the metrization. For details we refer to [49], ch. V. Let $\mathcal{L}$ be a line sheaf on X with admissible metric. On the finite R-modules $H^i(X,\mathcal{L})$, $i = 0,1$, one has derived volume forms. This is sufficient to define the *relative* EULER *characteristic* by

$$\chi_{X/Y}(\mathcal{L}) = \chi(\mathcal{L}) = \chi(H^0(X,\mathcal{L}),R) - \chi(H^1(X,\mathcal{L}),R).$$

The analogon of arithmetic genus is $\chi(\mathcal{O}_X)$.

Now we come to the local invariants δ_v for finite places $v \in M_K^{\mathrm{fin}}$. As usually we assume that X/Y is semistable. There are at most finitely many fibres X_v, which are not singular. The intersection points of the irreducible components of X_v are the singularities of this fibre. The cardinality of the residue field $k_v = R_v/m_v, m_v$ the maximal ideal, is denoted by q_v. Then we set

$$\delta_v = \sharp\{\text{singularities in } X_v\} \cdot \log q_v. \tag{6.11}$$

Theorem 6.19 (FALTINGS-NOETHER formula, [21], see also [54]). *With the above notations it holds that*

$$\chi(\mathcal{O}_X) = \frac{1}{12}(\omega_{X/Y} \cdot \omega_{X/Y}) + \sum_{v \in M_K} \delta_v \tag{6.12}$$

The gap is to explain δ_v for $v \in M_K^{\infty}$. We first give the following

Example 6.20 ([21]). Let E/K be an elliptic curve, say

$$E_v(\mathbb{C}) \cong \mathbb{C}/\mathbb{Z} + \mathbb{Z}\tau_v, \; v \in M_K^{\infty}, \; \tau_v \in \mathbb{H}.$$

Then the contributions at infinity in (6.12) are

$$\delta_v = -\varepsilon_v \cdot \log\{(2\pi)^{12} \cdot (\operatorname{Im}\tau_v)^6 \mid \triangle(\tau_v) \mid\}.$$

We used some notations appearing already in (6.12). In the general case of higher genus one needs Theta divisors and special values of Theta constants in order to determine δ_v, see [21]. In order to be quite explicit one needs an explicit knowledge of GREEN functions. For the case of genus 2 we refer to [12].

6.4 Conjectures of VOJTA and PARSHIN's Problem

6.21 PARSHIN**'s problem** [59]. Are there universal constants a_i, $i = 0, 1, 2$, a_0 depending on the genus g, such that for all semistable arithmetic surfaces X/Y it holds that

$$(\omega_{X/Y}^2) \leq a_2 \sum_{v \in M_K} \delta_v + a_1(2g-2)[K:\mathbb{Q}]d(K) + a_0 \ ? \tag{6.13}$$

At this place, we introduce the *logarithmic discriminant*

$$d(K) = \log |D_{K/\mathbb{Q}}| / [K:\mathbb{Q}] \tag{6.14}$$

of the number field K. The *relative logarithmic discriminant* of finite field extensions F/K is defined by

$$d(F/K) = \log |D_{F/K}| / [F:K], \ |D_{F/K}| := \sqrt[[K:\mathbb{Q}]]{N_{K/\mathbb{Q}}(D_{F/K})}, \tag{6.15}$$

with the relative discriminant $D_{F/K}$. It holds that

$$d(F) = d(K) + d(F/K). \tag{6.16}$$

This follows from

$$D_{F/\mathbb{Q}} = N_{K/\mathbb{Q}}(D_{F/K}) \cdot D_{K/\mathbb{Q}}^{[F:K]}$$

by taking logarithms and dividing by $[F:\mathbb{Q}] = [F:K] \cdot [K:\mathbb{Q}]$.

The inequality (6.13) is an arithmetic surface analogon of the BOGOMOLOV-MIYAOKA-YAU inequality. The latter is valid for smooth complex compact algebraic surfaces of general type.

We recall three of VOJTA's conjectures from [89] together with their connection with the Asymptotic FERMAT Theorem. Then we show that everything can be reduced (or transferred) to arithmetic surfaces of KODAIRA-PICARD type. Two of the conjectures involve WEIL heights, whereas $(\mathrm{Voj})_{\mathrm{Fal}}$ works with FALTINGS height. The latter is understood as an analogon near to PARSHIN's problem.

(Voj) $$h_{\omega_C} \leq \frac{a_1}{6} \ (20g-15) \ d(K(P)) + a_2$$

for smooth curves C/K of genus $g \geq 2$, $P \in C(\overline{\mathbb{Q}})$, constants $a_1, a_2 > 0$ depending on C, K and $[K(P):\mathbb{Q}]$.

($(\mathrm{Voj})_\varepsilon$) $$h_{\omega_C}(P) \leq (8g-4+\varepsilon)d(K(P)) + O(1)$$

for a given $\varepsilon > 0$, $P \in C(\overline{\mathbb{Q}})$, $O(1)$ depending on C, K, ε, C/K as in (Voj).

$(\mathrm{Voj})_{\mathrm{Fal}}$ $$h_{\mathrm{Fal}}(X/\mathcal{O}_F)/[F:\mathbb{Q}] \leq (\frac{1}{2}g + \varepsilon)d(F) + O(1)$$

for given $g \geq 2$, $\varepsilon > 0$, base $Y = \operatorname{Spec} R$, a finite set $S \subseteq M_K^{\mathrm{fin}}$, all finite extensions F/K and all semistable arithmetic surfaces $X/\mathcal{O}_F$ with good reduction outside of the places lying over S and general fibre X_F of genus g.

6.22 (VOJTA [89]). *The following implications hold:*

(i) PARSHIN's inequality (6.13) $\Longrightarrow$ VOJTA-conjecture (Voj);
(ii) VOJTA's conjecture (Voj) $\Longrightarrow$ Asymptotic FERMAT Theorem.

In the proofs one observes that not the full assumptions of the implications are needed but their restriction to special classes of arithmetic surfaces or algebraic curves, respectively. That's the point of change to the special classes introduced in Definitions 6.2. The main goal of this chapter is to verify the following:

Theorem 6.23. *The following restricted implications hold:*

(i) *If* PARSHIN*'s inequality* (6.13) *holds for all arithmetic surfaces of* KODAIRA-PICARD *type, then the* VOJTA *conjecture (Voj) holds for all* PICARD *curves.*
(ii) IF THE VOJTA'S CONJECTURE (VOJ) IS TRUE FOR ALL PICARD CURVES, THEN THE ASYMPTOTIC FERMAT THEOREM HOLDS.

Proof of (i): The first implication comes out directly from VOJTA's proof of the general implication 6.22 (i) as follows. We refer the reader to [84], pp. 164–173. The main tool is the construction of KODAIRA-PARSHIN covers of arithmetic surfaces. We denote by $KP(g)$ the class of all possible KODAIRA-PARSHIN covers of the class $Cl(g)$ of curves of genus g defined over any number field. The corresponding classes of arithmetic surface are denoted by $\mathbf{KP}(g)$ or $\mathbf{Cl}(g)$, respectively. Then one observes that VOJTA proved more precisely the restricted implications:

6.24. PARSHIN's inequality holds for $\mathbf{Cl}(4g-2) \Longrightarrow$

VOJTA's conjecture (Voj) for $Cl(g)$;

PARSHIN's inequality holds for $\mathbf{KP}(g) \Longrightarrow$

VOJTA's conjecture (Voj) for $Cl(g)$.

For $g = 3$ one gets the special restriction:

6.25. PARSHIN's inequality holds for $\mathbf{KP}(3) \Longrightarrow$

VOJTA's conjecture (Voj) for $Cl(3)$.

The further restriction to the subclass PIC $\subset Cl(3)$ of smooth PICARD curves on the right-hand side yields the first implication of Theorem 6.23. □

For the proof of the second implication we need intermediately the following section.

6.5 KUMMER Maps

Definitions 6.26. Let $F = (U_0, U_1, \ldots, U_q) : \mathbb{A}^{p+1} \longrightarrow \mathbb{A}^{q+1}$ be a rational map of affine spaces, $U_i = U_i(X_0, X_1, \ldots, X_p)$ polynomials with coefficients in the ring $\mathcal{O} = \mathcal{O}_M$ of a number field M. The map F is called $(\mathcal{O})$-*primitive*, if it maps $\mathcal{O}$-primitive points of $\mathbb{A}^{p+1}$ to $\mathcal{O}$-primitive points of $\mathbb{A}^{q+1}$. (Primitivity of points in $\mathbb{A}^r(\mathcal{O})$ means: the coordinates have no common prime divisor.) The map F is called *monomial*, if all polynomials U_i, $i = 0, \ldots, q$, are monomial. We call F *homogeneous*, if all polynomials U_i are homogeneous of the same degree.

Let n, k be natural numbers. An *affine* KUMMER *map* is a map

$$f : \mathbb{A}^{p+1}(\overline{\mathbb{Q}}) \dashrightarrow \mathbb{A}^{q+1}(\overline{\mathbb{Q}})$$
$$A = (a_0, a_1, \ldots, a_p) \mapsto U(A)^{n/k} := (U_0(A)^{n/k}, \ldots, U_q(A)^{n/k})$$

induced by a rational affine morphism $F = (U_0, \ldots, U_q)$ over $\mathcal{O}$ as described above. This map is finitely multivalued because of different possibilities to choose k-th roots; $U(A)^{n/k}$ denotes one of the possible $(q+1)$-tuples.

Now assume that F is homogeneous. Then it induces a rational map $\mathbb{P}(F) : \mathbb{P}^p \longrightarrow \mathbb{P}^q$ and a (multivalued) map

$$\mathbb{P}(f) : \mathbb{P}^p(\overline{\mathbb{Q}}) \dashrightarrow \mathbb{P}^q(\overline{\mathbb{Q}})$$
$$a = (a_0 : \ldots : a_p) \mapsto u(a)^{n/k} := (u_0^{n/k} : \ldots : u_q^{n/k}),\ u_i = U_i(a_0, \ldots, a_p).$$

We call $\mathbb{P}(F)$ a *projective* KUMMER *map*. It is said to be *monomial* or *primitive*, respectively, if F is. The rational number n/k is called the *weight* of the KUMMER map.

Proposition 6.27 (see SERRE, [70], 1.2, Corollary of Proposition 2). *Let L be a number field of degree $m = [L : \mathbb{Q}]$ and $|D_{L/\mathbb{Q}}| = \prod_p p_p^{\mu}$ the decomposition of the absolute discriminant into prime factors. Then one has the following estimation:*

$$\mu_p \leq (m-1) + m \cdot \log m / \log p. \tag{6.17}$$

Corollary 6.28. *Let $\mathbb{P}(f)$ be a projective monomial* KUMMER *map of weight n/k induced by F defined over $\mathbb{Z}$, $A \in \mathbb{A}^{p+1}(\mathbb{Z})$, $a = \mathbb{P}(A)$, $P = \mathbb{P}(f(a)) = (u_0 : u_1, \ldots : u_q)^{n/k}$, $u_i = U_i(A) \in \mathbb{Z}$, $u_i \neq 0, \pm 1$, an image point and $\mathbb{Q}(P)$ the definition field of P. Then for the logarithmic discriminant one has the following bound:*

$$d(\mathbb{Q}(P)) \leq (q+2)(1 + \log[\mathbb{Q}(P) : \mathbb{Q}]) \log \max_i(|u_i|). \tag{6.18}$$

Proof: We first remark that $\mathbb{Q}(P) \subseteq \mathbb{Q}(\sqrt[k]{u_0}, \ldots, \sqrt[k]{u_q})$, hence

$$p \mid D \Longrightarrow p = 2 \text{ or } p \mid u_0 \text{ or } \ldots \text{ or } p \mid u_q \tag{6.19}$$

For abbreviations we set $L = \mathbb{Q}(P)$, $D = D_{L/\mathbb{Q}}$, $m = [L : \mathbb{Q}]$. From (6.17) we get

$$\log |D| = \sum_p \mu_p \cdot \log p \leq \sum_{p|D} (m \cdot \log p + m \cdot \log m).$$

Dividing by m we obtain

$$\begin{aligned} d(L) &\leq \sum_{p|D} \log p + \sum_{p|D} \log m = \sum_{p|D} \log p + (\log m) \sum_{p|D} 1 \\ &\leq \sum_{p|D} \log p + (\log m) \sum_{p|D} \log p = (1 + \log m) \sum_{p|D} \log p \\ &\leq (1 + \log m)(\log 2 + \sum_{p|u_0} logp + \ldots + \sum_{p|u_q} logp) \\ &\leq (q+2)(1 + \log m) \log \max_i (|u_i|) \end{aligned}$$

In the last step we used $u_i \neq 0$, ± 1, and in the step before the implication (6.19). □

6.6 Proof of the Main Implication

The Main Implication is

6.29. *If the* PARSHIN *inequality* (6.13) *holds for all arithmetic surfaces of* KODAIRA-PICARD *type, then the Asymptotic* FERMAT *Theorem holds.*

Looking back to Theorem 6.23 it remains to carry out the

Proof of the implication 6.23 (ii): Let C/K be a smooth curve of genus 3. If two heights h, h' on $C(\overline{\mathbb{Q}})$ differ only by a real bounded function, then we write $h \asymp h'$. The canonical map of C embeds C into $\mathbb{P}^2$, and the pullback of the very ample sheaf $\mathcal{O}_{\mathbb{P}^2}(1)$ to C is the canonical sheaf on C. We set $h = h_{\mathbb{P}^2 | C(\overline{\mathbb{Q}})}$, where $h_{\mathbb{P}^2}$ denotes WEIL's coordinate height function on the projective plane introduced in 6.5.

Take an arithmetic surface model X/R of C/K and let h_{Ar} be the ARAKELOV height function introduced in 6.16. The Corollary 6.17 and the functorial property (func), see 6.3, are brought together in the following

Lemma 6.30. On smooth curves C of genus 3 defined over $\overline{\mathbb{Q}}$, especially on smooth PICARD curves, it holds that $h \asymp h_{\omega_C} \asymp h_{Ar}$.

Starting with VOJTA's conjecture (Voj) we can find constants $b, c_1, c_2 > 0$ such that for all $P \in C(\overline{\mathbb{Q}}), C$ a fixed curve of genus 3, we have with regard to Lemma 6.30 two inequalities

$$h(P) \leq h_{\omega_C}(P) + b \leq c_1 d(K(P)) + c_2. \tag{6.20}$$

Let $P = (u : v : w)^{n/k}$ be a point of $\mathbb{P}^2$ with primitive $\mathbb{Z}$-coordinates u, v, w. We calculate $h(P) = h_{\mathbb{P}^2}(P)$ by means of (6.5). The finite part is 0 because (u, v, w) is primitive. Working in the number field $E = \mathbb{Q}(\sqrt[k]{u}, \sqrt[k]{v}, \sqrt[k]{w})$ of degree l, say, we get

$$h(P) = \frac{n}{[E:\mathbb{Q}]} \sum_{v \in M_E} \log \max\{|u|^{\varepsilon_v/l}, |v|^{\varepsilon_v/l}, |w|^{\varepsilon_v/l}\},$$

$$h(P) = \frac{n}{k} \log \max\{|u|, |v|, |w|\}. \tag{6.21}$$

We know smooth PICARD curves C, which allow a simple transfer of $\mathbb{Z}$-solutions of the FERMAT equations to algebraic points of C.

Take for example

$$C : ZY^3 = X^4 + Z^4. \tag{6.22}$$

For the transfer mentioned above we use the following projective monomial primitive KUMMER maps from $\mathbb{P}^2(\overline{\mathbb{Q}})$ to $\mathbb{P}^2(\overline{\mathbb{Q}})$ defined over $\mathbb{Z}$ of weight $n/12, n \in \mathbb{N}$:

$$\varphi_n : (a : b : c) \mid - - - \longrightarrow (u : v : w)^{n/12} = (a^3 b : c^4 : b^4)\ . \tag{6.23}$$

Assume that $(a : b : c)$ lies on the FERMAT *curve* $C_n : X^n + Y^n = Z^z$. Then it is easy to see that any φ_n-image lies on C. Putting all KUMMER maps φ_n together, we obtain the finitely multivalued map

$$\Phi : \{\mathbb{Z} - \text{solutions of FERMAT equations}\} - - - \longrightarrow C(\overline{\mathbb{Q}}).$$

Now assume that $(a : b : c)$ is a non-trivial $\mathbb{Z}$-point on $C_n, (a, b, c)$ a primitive $\mathbb{Z}$-triple, $a, b, c \neq 0$. We fix a φ_n-image $P = (u : v : w)^{n/k}$ on $C(\overline{\mathbb{Q}})$; $k = 12$ is not important. The assumptions for applying (6.21) and Corollary 6.28 are satisfied. From the corollary we obtain

$$d(\mathbb{Q}(P)) \leq (q+2)(1 + \log[\mathbb{Q}(P) : \mathbb{Q}]) \log \max\{|u|, |v|, |w|\} \tag{6.24}$$

($q = 2$ is not important). The field $\mathbb{Q}(\sqrt[k]{a}, \sqrt[k]{b}, \sqrt[k]{c})$ contains $\mathbb{Q}(P)$. Therefore $m = [\mathbb{Q}(P) : \mathbb{Q}] \leq 3k$, hence

$$d(\mathbb{Q}(P)) \leq 4(q+2)(1 + \log 3k) \log \max\{|u|, |v|, |w|\}. \tag{6.25}$$

Now put together the estimations (6.20), (6.25) and the identity (6.21). Since C is defined over $\mathbb{Q}$, we have $K(P) = \mathbb{Q}(P)$. Altogether one gets

$$\frac{n}{k} \log \max\{|u|, |v|, |w|\} \leq c_1(q+2)(1 + \log 3k) \log \max\{|u|, |v|, |w|\} + c_2$$

or

$$\left[\frac{n}{k} - c_1(q+2)(1 + \log 3k)\right] \log \max\{|u|, |v|, |w|\} \leq c_2. \tag{6.26}$$

By the assumption of non-triviality of the solution $(a : b : c)$ of the n-th FERMAT equation, in the correspondence (6.23) we are sure that $\log \max\{|u|, |v|, |w|\} > 1$. Therefore we can omit this logarithmic factor in (6.26). Finally we see that

$$n \leq kc_1(q+2)(1 + \log 3k) + kc_2. \tag{6.27}$$

This means that the FERMAT equation of degree n has only trivial $\mathbb{Z}$-solutions, if n is greater than the right side of the inequality (6.27). So the Asymptotic FERMAT Theorem can be derived from the restricted PARSHIN problem described in 6.25, if it has an affirmative answer. This completes the proof of Theorem 6.23 and of its composed Main Implication. □

Remark 6.31. VOJTA's conjecture (Voj) works with WEIL heights on curves. Lemma 6.30 teaches us that in the case of genus 3 curves we can change over to the ARAKELOV heights of points coming from intersection theory on arithmetic surfaces. So the whole proof of the Main Implication goes through using "only" the geometry on arithmetic surfaces.

At the end of this section we turn to the other VOJTA conjectures.

Theorem 6.32 (VOJTA, [89]). *The following implications hold:*

(i) VOJTA-conjecture $(\mathrm{Voj})_{\mathrm{Fal}} \Longrightarrow$ VOJTA-conjecture $(\mathrm{Voj})_\varepsilon$;
(ii) VOJTA conjecture $(\mathrm{Voj})_\varepsilon \Longrightarrow$ Asymptotic FERMAT Theorem.

As in the proof of Theorem 6.22, VOJTA uses the KODAIRA-PARSHIN construction of arithmetic surface coverings to prove the first implication (i) of Theorem 6.32. As before, this proof can be restricted to arithmetic surfaces of KODAIRA-PICARD type in the assumption and arithmetic surfaces of PICARD type in the conclusion of (i), respectively. Now observe that conjecture $(\mathrm{Voj})_\varepsilon$ is a little bit weaker than (Voj). Composing with implication 6.23 (ii) we obtain the final

Corollary 6.33. *The following restricted implications hold:*

(i) *If the* VOJTA *conjecture* $(\mathrm{Voj})_{\mathrm{Fal}}$ *is true for all arithmetic surfaces of* KODAIRA-PICARD *type, then the* VOJTA *conjecture* $(\mathrm{Voj})_\varepsilon$ *holds for all* PICARD *curves.*
(ii) *If the* VOJTA *conjecture* $(\mathrm{Voj})_\varepsilon$ *is true for all* PICARD *curves, then the Asymptotic* FERMAT *Theorem holds.*

Remark 6.34. Some preliminary work on the estimation of constants occurring in the conjectures and conclusions has been already done in [89]. But more effort is needed on the way to effectiveness.

The hard kernel of the proof of the Main Implication 6.29 is VOJTA's proof of 6.22 (i) in [89]. There one needs the whole machinery of ARAKELOV theory for arithmetic surfaces from which we only introduced the basic tools in order to encourage the reader to intervene there.

It seems to be hopeless now to check directly the conjecture (Voj) for PICARD curves, even for the special one C defined in (6.22). With regard to the KUMMER transfer Φ the most interesting solutions have coefficients in KUMMER extensions L of $\mathbb{Q}$ generated by three twelfth roots of natural numbers. Fixing one KUMMER field L the set $C(L)$ of L-solutions of (6.22) is finite by the MORDELL-FALTINGS Theorem. But until now there is no method to find these solutions effectively. So a case by case checking of (Voj) for special PICARD curves is as far of our reach as the solution of the FERMAT problem itself because the solutions of the FERMAT equations are discovered among the solution sets $C(L)$, where L runs through the KUMMER fields described above.

On the other hand our inefficiency for checking (Voj) even for simple PICARD curves emphasizes the meaning of PARSHIN's analogon of the BOGOMOLOV-MIYAOKA-YAU inequality. So it is important to check it for special curves. In this sense the following problems seem to be a good orientation for further progress in the theory of diophantine equations:

6.35 Problem. *Check* PARSHIN*'s problem* 6.21 *for* PICARD *curves.*

Hint. Use all informations around HILBERT's Twelfth Problem connected with the ball uniformizing the moduli space of PICARD curves by means of PICARD modular theta constants. It seems to be useful to transfer the methods of BOST, MESTRE, MORET-BAILLY used in [13] for the calculation of invariants of arithmetic surfaces of genus 2 to PICARD curves.

6.36 Problem. *Find the moduli space of* KODAIRA-PICARD *curves of biquadratic type, a nice uniformization of suitable (quasi)compactification, curve equations and, moreover, as much knowledge as possible in analogy to the theory of* PICARD *curves.*

7 Appendix I
A Finiteness Theorem for Picard Curves With Good Reduction
(by J. Estrada-Sarlabous)

In this appendix we study the PICARD curves defined over the spectrum of a DEDEKIND domain and obtain affine models (Lemma 7.3), normal forms (Remark 7.7) and a characterization of the minimal normal forms (Lemma 7.9).

Inspired in the treatment of the elliptic curves exposed by HUSEMÖLLER in [41], we obtain a projective isomorphism classification of the PICARD curves in characteristic $p > 3$ (§ 1.5) and an alternative proof of the conjecture of SHAFAREVIČ [103] on the finiteness of the number of isomorphism classes of curves with good reduction outside a finite set of discrete valuations of a number field for the case of PICARD curves (Theorem 7.12), which is based on SIEGEL's Theorem on the finiteness of integral points on curves.

This alternative proof reduces the effectivity of finding the isomorphism classes of PICARD curves with good reduction outside a finite set of discrete valuations to the effectivity of finding the integral points of elliptic curves defined over the ring of integers of the number field.

Due to the constructive nature of this proof, it is of theoretical interest, besides the quite general finiteness theorems of G. FALTINGS (see [FW] and [Lan]).

This paper is a revised and improved version of a talk given at the I. Symposium on the Development of Mathematics, Havana'90.

7.1 Some Definitions and Known Results

By a curve of genus g over a scheme S we mean a smooth projective morphism $p : \mathcal{C} \to S$ whose geometric fibres are irreducible curves of genus g. In particular, a curve of genus 0 over S is called a twisted $\mathbb{P}^1_S$, [101].

Let D be a DEDEKIND domain, K its fields of fractions and $\mathcal{C}$ a complete, smooth and absolutely irreducible curve over K of genus g.

Definition 7.1. A curve of genus g, $\mathcal{C} \to S$ is called *n-gonal cyclic* if there exists a nontrivial automorphism $\sigma \in \mathrm{Aut}_S(\mathcal{C})$ such that σ has order n (i.e., $\sigma^n = identity$) and $\mathcal{C}/ < \sigma >$ is a twisted $\mathbb{P}^1_S$ (compare with [Est]).

Remark 7.2. *For all subgroups $G \subseteq \mathrm{Aut}_S(\mathcal{C})$, $\mathcal{C}/G$ is a curve over S. If $\mathcal{C}$ is a n-gonal cyclic curve over S, the canonical projection $\pi : \mathcal{C} \to \mathcal{C}/\langle\sigma\rangle \cong \mathbb{P}^1_S$ is finite surjective S-morphism of degree n.*

7.2 Affine Models of n-gonal Cyclic Curves

Let $\overline{K}$ be an algebraic closed field containing K and let $\mathcal{C}$ be a curve defined over Spec(K), such that there exists a nontrivial automorphism $\sigma \in \mathrm{Aut}_K(\mathcal{C})$ with $\sigma^n = identity$, n prime and $(\mathcal{C} \otimes_K \overline{K})/\langle\sigma\rangle \cong \mathbb{P}^1_{\overline{K}}$.

Then the field of rational functions of C, $R(\mathcal{C})$, is a cyclic GALOIS extension of the field of rational functions of $\mathbb{P}^1_{\overline{K}}$, $\overline{K}(x)$, therefore there exists an element $y \in R(\mathcal{C})$ such that $\sigma(y) = \zeta y$, $R(\mathcal{C}) = \overline{K}(y)$ and $y^n \in \overline{K}(x)$ for ζ primitive n-th root of unity. So, $\mathcal{C}$ has the affine model

$$\mathcal{C}: \quad y^n = a_0(x-a_1)^{m_1}(x-a_2)^{m_2}\dots(x-a_s)^{m_s} \tag{7.1}$$

with $m_j \in \mathbb{Z}$, $a_j \in \overline{K}$, $a_i \neq a_j$ for $i \neq j$.

Without loss of generality we may assume that

$$n \nmid m_1 + m_2 + \dots + m_s \ , \tag{7.2}$$

otherwise if $m_1 + m_2 + \dots + m_s = nq$, for $X = (x-a_s)^{-1}$, $Y = y/(x-a_s)^q$ we obtain $Y^n = A_0(X-B_1)^{m_1}(-B_2)^{m_2}\dots(X-B_{s-1})^{m_{s-1}}$ with $B_j = (a_j - a_s)^{-1}$, $A_0 = a_0 \prod_{0<j<s} (a_s - a_j)^{m_j}$, $R(\mathcal{C}) = \overline{K}(X,Y)$, $\sigma(Y) = \zeta Y$ and we can further analogously proceed, until we reach the first j such that $n \nmid m_1 + m_2 + \dots + m_{s-1}$.

We can also assume that

$$n > m_j > 0 \ . \tag{7.3}$$

In fact, if $M_j = nq_j + r_j$, $0 \leq r_j < n$, put $Y = y(x-a_j)^{-q_j}$ and $X = x$, so $Y^n = a_0(X-a_1)^{m_1}\dots(X-a_j)^{r_j}\dots(X-a_s)^{m_s}$, and $R(\mathcal{C}) = \overline{K}(X,Y)$.

By means of the RIEMANN-HURWITZ formula, for char$(\overline{K}) \neq n$ we can compute the genus of the curve C with affine model (7.1),

$$g = (n-1)(s-1)/2 \tag{7.4}$$

in the nonsingular case $m_j = 1$, $1 \leq j \leq s$.

In particular, if $n = g = 3$, we get $s = 4$ and there are only four possible cases under the assumptions (7.2), (7.3):

$$\begin{array}{ll}
\text{(a)} & y^3 = a_0(x-a_1)(x-a_2)(x-a_3)(x-a_4) \\
\text{(b)} & y^3 = a_0(x-a_1)(x-a_2)(x-a_3)(x-a_4)^2 \\
\text{(c)} & y^3 = a_0(x-a_1)(x-a_2)^2(x-a_3)^2(x-a_4)^2 \\
\text{(d)} & y^3 = a_0(x-a_1)^2(x-a_2)^2(x-a_3)^2(x-a_4)^2
\end{array}$$

with $a_j \in \overline{K}$.

Lemma 7.3. *The four cases (a)–(d) above are birationally equivalent.*

Proof: (a) and (b) are birationally equivalent: In fact, put in (b) $y^3 = a_0(x - a_1)(x - a_2)(x - a_3)(x - a_4)^2$, $Y = y/(x - a_4)^2$, $X = (x - a_4)^{-1}$ so we obtain (a) $Y^3 = A_0(X - A_1)(X - A_2)(X - A_3)(X - A_4)$ with $A_0 = a_0 \prod_{0<j<4} (a_4 - a_j)$, $A_1 = 0$, $A_j = (a_{j-1} - a_4)^{-1}$ for $j > 1$.

(c) and (d) are birationally equivalent: Put in (c) $y^3 = a_0(x - a_1)(x - a_2)^2(x - a_3)^2(x - a_4)^2$, $Y = y/(x - a_1)^3$, $X = (x - a_1)^{-1}$, so we obtain (d) $Y^3 = A_0(x - A_1)^2(x - A_2)^2(x - A_3)^2(x - A_4)^2$ with $A_0 = a_0 \prod_{j=2}^{4} (a_j - a_1)^2$, $A_1 = 0$, $A_j = (a_{j-1} - a_1)^{-1}$ for $j > 1$. Squaring (d), we get $y^6 = a_0^2(x - a_1)^4(x - a_2)^4(x - a_3)^4(x - a_4)^4$. Put $X = x$ and $Y = y^2 \prod_{0<j<5} (x - a_i)^{-1}$, then we obtain (a) $Y^3 = A_0(X - A_1)(X - A_2)(X - A_3)(X - A_4)$ with $A_0 = a_0^2$, $A_j = a_j$, $j > 1$. Reciprocally, starting from (a), for $X = x$ and $Y = y^2$ we see that $Y^3 = a_0^2 \prod_{0<j<5} (x - a_j)^2$ so, (d) $Y^3 = A_0 \prod_{0<j<5} (X - A_j)^2$ holds for $A_0 = a_0^2$, $A_j = a_j$, $j > 1$. Hence (a) and (d) are birationally equivalent too. □

7.3 Normal Forms of PICARD Curves

Definition 7.4. *A* PICARD *curve over* $S = \mathrm{Spec}(D)$ is a trigonal cyclic curve of genus 3 over S, $K = Quot(D)$.

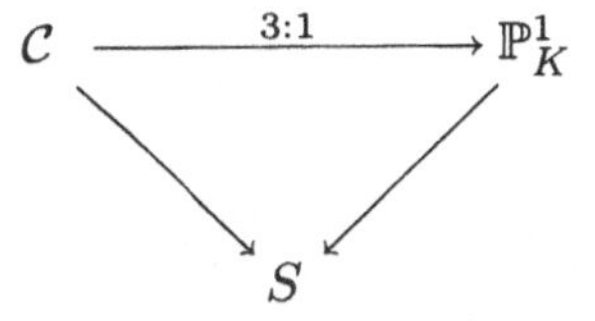

After Lemma 7.3 above, the general fibre C_K of a PICARD curve over S has a projective model

$$C_K : f(x, y, w) = wy^3 - a_0 \prod_{0<i<5} (x - a_i w) = wy^3 - w^4\, p_4\!\left(\frac{x}{w}\right) = 0, \qquad (7.5)$$

with $p_4(t) = \alpha_4 t^4 + \alpha_3 t^3 + \alpha_2 t^2 + \alpha_1 t + \alpha_0$, $\alpha_j, a_j \in \overline{K} \supseteq K$, $\alpha_4 \neq 0$, $[\overline{K} : K] < \infty$.

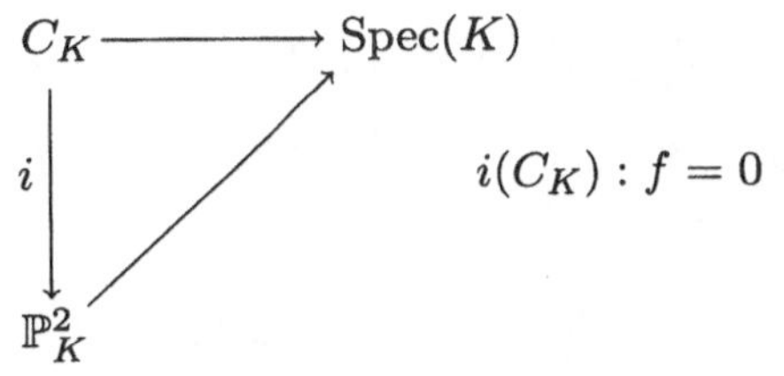

We call the equation (7.5) a *separated form of the* PICARD *curve.* The K-automorphisms of $\mathbb{P}^2_K$ which preserve the separated form of the PICARD curve (i.e., such that the diagram

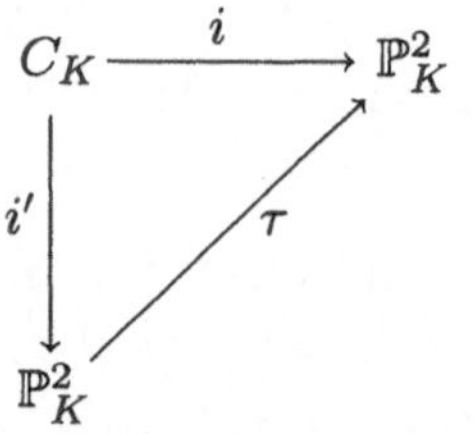

commutes, i, i' separated form embeddings) are special TSCHIRNHAUS transformations (see [33]) which can be represented by the elements $\tau \in \mathbb{PG}l_3(K) \cong \mathrm{Aut}(\mathbb{P}^2_K)$ of type

$$\tau = \begin{pmatrix} 1 & 0 & 0 \\ r & u^3 & 0 \\ 0 & 0 & u^4 \end{pmatrix} \quad \text{with } u, r \in K,\ u \neq 0 \tag{7.6}$$

and the corresponding change of variables is

$$\begin{pmatrix} w \\ x \\ y \end{pmatrix} = \tau \begin{pmatrix} \bar{w} \\ \bar{x} \\ \bar{y} \end{pmatrix}.$$

The substitution by a special TSCHIRNHAUS transformation (7.6) into the separated form (7.5) yields a new separated form in terms of the variables $\bar{w}$, $\bar{x}$, $\bar{y}$:

$$\bar{f} = \bar{w}\bar{y}^3 - (\bar{\alpha}_4\bar{x}^4 + \bar{\alpha}_3\bar{w}\bar{x}^3 + \bar{\alpha}_2\bar{x}^2\bar{w}^2 + \bar{\alpha}_1\bar{x}\bar{w}^3 + \bar{\alpha}_0\bar{w}^4) = \bar{w}\bar{y}^3 - \bar{w}^4\bar{p}_4(\bar{x}/\bar{w})$$

with

$$\begin{cases} \bar{\alpha}_4 = \alpha_4 \\ u^3\bar{\alpha}_3 = \alpha_3 + 4r\alpha_4 \\ u^6\bar{\alpha}_2 = \alpha_2 + 3r\alpha_3 + 6r^2\alpha_4 \\ u^9\bar{\alpha}_1 = \alpha_1 + 2\alpha_2 r + 3\alpha_3 r^2 + 4r^3\alpha_4 \\ u^{12}\bar{\alpha}_0 = \alpha_0 + \alpha_1 r + \alpha_2 r^2 + \alpha_3 r^3 + \alpha_4 r^4\,. \end{cases} \tag{7.7}$$

For $\mathrm{char}(K) \neq 2, 3$ we consider the following parameters of the PICARD curve:

$$i = (12\alpha_0\alpha_4 - 3\alpha_1\alpha_3 + \alpha_2^2)/6 = u^{12}\,\bar{i} \tag{7.8}$$

$$j = (72\alpha_0\alpha_2\alpha_4 + 9\alpha_1\alpha_2\alpha_3 - 2\alpha_2^3 - 27\alpha_1^2\alpha_4 - 27\alpha_0\alpha_3^2)/72 = u^{18}\bar{j} \tag{7.9}$$

$$i_\alpha = i^3/j^2 = \bar{i}_\alpha \tag{7.10}$$

$$\Delta = (i^3 - 6j^2)/27 = j^2(i_\alpha - 6)/27 = u^{36}\bar{\Delta} \tag{7.11}$$

These parameters are related to the invariants of the fourth degree binary form $w^4 p_4(x/w)$ with respect to the action of $\mathbb{G}l_2(K)$, for $\mathrm{char}(K) \neq 2, 3$ (see for example [Cle] or [Dix]) and they can be expressed in terms of the roots of the polynomial $p_4(x) = a_0 \prod_{0<i<5} (x - a_i)$ by the following formulae:

Set $u = (a_1 - a_2)(a_4 - a_3)$, $v = (a_1 - a_4)(a_3 - a_2)$, $w = (a_1 - a_3)(a_2 - a_4)$, then $u + v + w = 0$ and for $\epsilon = -v/w$ it holds

$$
\begin{aligned}
i &= a_0^2(u^2 + v^2 + w^2)/12 \\
j &= a_0^3(u - v)(v - w)(w - u)/72 \\
i_\alpha &= \frac{i^3}{j^2} = \frac{3(u^2 + v^2 + w^2)^3}{(u - v)^2(u - w)^2(v - w)^2} = \frac{24(1 - \epsilon + \epsilon^2)^3}{(1 + \epsilon)^2(2 - \epsilon)^2(1 - 2\epsilon)^2} \qquad (7.12) \\
\Delta &= a_0^6 u^2 v^2 w^2
\end{aligned}
$$

From the formulae (7.9)–(7.12) it is easy to deduce that "$p_4(x)$ has double root" is equivalent to $\Delta = 0$.

Lemma 7.5. *For any* PICARD *curve over* $S = \mathrm{Spec}(D)$ *with separate form of type (7.5) and coefficients in* K *there exists a separate form with all its coefficients and parameters in* D.

Proof: Take a projective model of the generic fibre C_K,

$$C_K: \quad \bar{f} = \bar{w}\bar{y}^3 - (\bar{\alpha}_4\bar{x}^4 + \bar{\alpha}_3\bar{w}\bar{x}^3 + \bar{\alpha}_2\bar{x}^2\bar{w}^2 + \bar{\alpha}_1\bar{x}\bar{w}^3 + \bar{\alpha}_0\bar{w}^4)$$

with $\bar{\alpha}_j \in K$ and $\bar{\alpha}_4 \neq 0$ and let u be a common denominator of all $\bar{\alpha}_j$, i.e, $u\bar{\alpha}_j \in D$. Then the diagonal TSCHIRNHAUS transformation $\tau = \mathrm{diag}(\bar{\alpha}^{-1}, u^3, u^4)$ provides us a separated form

$$f = wy^3 - (x^4 + \alpha_3 wx^3 + \alpha_2 x^2 w^2 + \alpha_1 xw^3 + \alpha_0 w^4)$$

with

$$
\begin{cases}
\alpha_3 = u^3\bar{\alpha}_3, \\
\alpha_2 = u^6\bar{\alpha}_2\bar{\alpha}_4, \\
\alpha_1 = u^9\bar{\alpha}_1\bar{\alpha}_4^2, \\
\alpha_0 = u^{12}\bar{\alpha}_0\bar{\alpha}_4^3\,.
\end{cases} \qquad (7.13)
$$

Thus f has all its coefficients and parameters in D. Furthermore, if $\mathrm{char}(K) \neq 2$, we can also assume that the coefficient α_3 is 0, in fact by means of TSCHIRNHAUS transformation

$$\begin{pmatrix} w \\ x \\ y \end{pmatrix} = \begin{pmatrix} 1 & 0 & 0 \\ r & 1 & 0 \\ 0 & 0 & 1 \end{pmatrix} \begin{pmatrix} w' \\ x' \\ y' \end{pmatrix}$$

with $r = -\alpha_3/4$, we can obtain a separated form

$$f' = w'y'^3 - (x'^4 + \alpha_2' x'^2 w'^2 + \alpha_1' x' w'^3 + \alpha_0' w'^4)$$

with $\alpha_2' = \alpha_2 - \frac{3\alpha_3^2}{8}$, $\alpha_1' = \alpha_1 - \frac{\alpha_2\alpha_3}{4} + \frac{\alpha_3^3}{8}$, $\alpha_0' = \alpha_0 - \frac{\alpha_3\alpha_1}{4} + \frac{\alpha_3^2\alpha_2}{16} - \frac{3\alpha_3^4}{256}$. □

Definition 7.6. A separated form (7.5) with $\alpha_4 = 1$, $\alpha_3 = 0$, $\alpha_2, \alpha_1, \alpha_0 \in D$ is called a *normal form.*

Remark 7.7. *If the generic fibre of the* PICARD *curve* C *over* Spec(D) *has a projective model*

$$f = wy^3 - (\alpha_4 x^4 + \alpha_3 wx^3 + \alpha_2 x^2 w^2 + \alpha_1 xw^3 + \alpha_0 w^4), \quad \alpha_j \in K,$$

then for char$(K) \neq 2$ *there exists a normal form for* C.

Any pair of normal forms of a PICARD curve $f = f(w,x,y)$, $\bar{f} = \bar{f}(\bar{w},\bar{x},\bar{y})$ are related by a diagonal TSCHIRNHAUS transformation

$$\begin{pmatrix} w \\ x \\ y \end{pmatrix} = \tau \begin{pmatrix} \bar{w} \\ \bar{x} \\ \bar{y} \end{pmatrix} \quad \text{with } \tau = \begin{pmatrix} 1 & 0 & 0 \\ 0 & u^3 & 0 \\ 0 & 0 & u^4 \end{pmatrix} \quad \text{and } u \in K .$$

The corresponding coefficients and parameters are related by the formulae:

$$\begin{cases} u^{-6}\alpha_2 = \bar{\alpha}_2, & u^{-9}\alpha_1 = \bar{\alpha}_1, & u^{-12}\alpha_0 = \bar{\alpha}_0, \\ \alpha_4 = \bar{\alpha}_4 = 1, & \alpha_3 = \bar{\alpha}_3 = 0, & u^{-12} i = \bar{i}, \\ u^{-18} j = \bar{j}, & u^{-36}\Delta = \bar{\Delta}, & i_\alpha = \bar{i}_\alpha . \end{cases} \tag{7.14}$$

7.4 Conditions for Smoothness

For

$$C_K : \ f = wy^3 - (x^4 + \alpha_2 x^2 w^2 + \alpha_1 xw^3 + \alpha_0 w^4), \quad \alpha_j \in K$$

the corresponding PICARD curve is smooth iff f, f_x, f_y and f have no common roots, i.e., if the system of equations

$$\begin{aligned} f &= wy^3 - (x^4 + \alpha_2 x^2 w^2 + \alpha_1 xw^3 + \alpha_0 w^4) = 0 \\ f_x &= -(4x^3 + 2\alpha_2 xw^2 + \alpha_1 w^3) = 0 \\ f_y &= 3y^2 w = 0 \\ f_w &= y^3 - (2\alpha_2 x^2 w + 3\alpha_1 xw^2 + 4\alpha_0 w^3) = 0 \end{aligned}$$

has no solution in K.

If char$(K) = 3$, the points $(w : x : y) = (1 : x_i : y_{j_i})$ are singular, where x_i are the solutions of $x^3 + 2\alpha_2 x + \alpha_1 = 0$ and y_{j_i} are the solutions of $y^3 = \alpha_0 + 2\alpha_2 x_i^2$.

If char$(K) \neq 3$, then any singular point has $w \neq 0$ (if $w = 0$, $f = 0$ implies $x = 0$ and from $f_w = 0$ we should obtain $y = 0$), thus $y = 0$ and from $f = f_x = y = 0$, we see that $p_4(t)$ and $p_4'(t)$ should have a common root, and this is equivalent to the vanishing of the discriminant Δ of p_4. If $\Delta = 0$, then the points $(1 : a : 0)$ are singular for each double root a of $p_4(t)$. So, if char$(K) \neq 3$, C_K is smooth iff $\Delta \neq 0$.

7.5 Projective Isomorphy Classification in Characteristic > 3

Suppose that C_K and $\bar{C}_K$ are the generic fibres of two PICARD curves with normal forms

$$C_K: \quad f = wy^3 - (x^4 + \alpha_2 x^2 w^2 + \alpha_1 x w^3 + \alpha_0 w^4) = 0,$$

$$\bar{C}_K: \quad \bar{f} = \bar{w}\bar{y}^3 - (\bar{x}^4 + \bar{\alpha}_2 \bar{x}^2 \bar{w}^2 + \bar{\alpha}_1 \bar{x} \bar{w}^3 + \bar{\alpha}_0 w^4) = 0,$$

and with the same absolute invariant i_α. If

$$\begin{array}{ccc} \tau : \bar{C}_K & \longrightarrow & C_K \\ & \searrow \quad \swarrow & \\ & \mathbb{P}^2_K & \end{array}$$

is a projective isomorphism, then

$$\begin{pmatrix} w \\ x \\ y \end{pmatrix} = \begin{pmatrix} 1 & 0 & 0 \\ 0 & u^3 & 0 \\ 0 & 0 & u^4 \end{pmatrix} \begin{pmatrix} \bar{w} \\ \bar{x} \\ \bar{y} \end{pmatrix} \quad \text{for } 0 \neq u \in K$$

and the relations among the coefficients and parameters 7.14 hold (in particular, $\alpha_j \neq 0$ iff $\bar{\alpha}_j \neq 0$).

We can study these relations in six disjoint cases:

Case 1:
$\alpha_1\alpha_2 \neq 0$. Then, C_K and $\bar{C}_K$ are projective isomorphic iff the quotient $(\alpha_1\bar{\alpha}_2)/(\bar{\alpha}_1\alpha_2)$ is a cube in K and the conditions $\alpha_1^2\bar{\alpha}_2^3 = \bar{\alpha}_1^2\alpha_2^3$, $\alpha_0\bar{\alpha}_2^2 = \bar{\alpha}_0\alpha_2^2$ are satisfied. Thus, in any field extension of K containing the cubic root of $(\alpha_1\bar{\alpha}_2)/(\bar{\alpha}_1\alpha_2)$, C_K and $\bar{C}_K$ are projective isomorphic iff $\alpha_1^2\bar{\alpha}_2^3 = \bar{\alpha}_1^2\alpha_2^3$ and $\alpha_0\bar{\alpha}_2^2 = \bar{\alpha}_0\alpha_2^2$.

Case 2:
$\alpha_1 = 0, \alpha_2 \neq 0$. Then C_K and $\bar{C}_K$ are projective isomorphic iff the quotient $\alpha_2/\bar{\alpha}_2$ is a 6-th power in K and $\alpha_0\bar{\alpha}_2^2 = \bar{\alpha}_0\alpha_2^2$. Thus, in any field extension of K containing the 6-th root of $\alpha_2/\bar{\alpha}_2$, C_K and $\bar{C}_K$ are projective isomorphic iff $\alpha_0\bar{\alpha}_2^2 = \bar{\alpha}_0\alpha_2^2$.

Case 3:
$\alpha_2 = 0, \alpha_1\alpha_0 \neq 0$. Here C_K and $\bar{C}_k$ are projective isomorphic iff the quotient $(\alpha_0\bar{\alpha}_1)/(\bar{\alpha}_0\alpha_1)$ is a cube in K and $\alpha_0^3\bar{\alpha}_1^4 = \bar{\alpha}_0^3\alpha_1^4$. Thus, in any field extension of K containing the cubic root of $(\alpha_0\bar{\alpha}_1)/(\bar{\alpha}_0\alpha_1)$, C_K and $\bar{C}_K$ are projective isomorphic iff $\alpha_0^3\bar{\alpha}_1^4 = \bar{\alpha}_0^3\alpha_1^4$.

Case 4:
$\alpha_1 \neq 0, \alpha_2 = \alpha_0 = 0$. Now C_K and $\bar{C}_K$ are projective isomorphic iff $\alpha_1/\bar{\alpha}_1$ is a nine-th power in K. Hence, C_K and $\bar{C}_K$ are projective isomorphic in any field extension of K containing a nine-th power of $\alpha_1/\bar{\alpha}_1$.

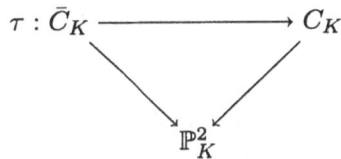

Case 5:
$\alpha_0 \neq 0, \alpha_2 = \alpha_1 = 0$. Then C_K and $\bar{C}_K$ are projective isomorphic iff $\alpha_0/\bar{\alpha}_0$ is a 12th power in K. Thus, C_K and $\bar{C}_K$ are projective isomorphic in any field extension of K containing a 12th power of $\alpha_0/\bar{\alpha}_0$.

Case 6:
$\alpha_2 = \alpha_1 = \alpha_0 = 0$. The projective isomorphism class consists of a single element with equation $wy^3 = x^4$.

7.6 Minimal Normal Forms for PICARD Curves

If D is a discrete valuation ring (i.e., principal with one nonzero prime ideal Dp) the order function ord_p is a valuation denoted by ν and $D = \{a \in K; \nu(a) \geq 0\}$.

Definition 7.8. Let K be a field with a discrete valuation ν and let $\mathcal{C}$ be a PICARD curve defined over Spec(D). A *minimal normal form* for $\mathcal{C}$ is a normal form with all its coefficients α_j in the valuation ring D of K such that $\nu(\Delta)$ is minimal among all such equations with coefficients in D.

Proposition 7.9. *If all the coefficients of a separated form are in D and $0 \leq \nu(\Delta) < 36$, then the separated form is a minimal normal form.*

Proof: For any normal form with coefficients $\alpha_j \in D$ in the variables w, x, y it holds $\nu(\Delta) \geq 0$. If $\nu(\Delta) \geq 36$, by means of TSCHIRNHAUS transformation $\tau = \text{diag}(1, u^3, u^4)$ we get a new normal form with coefficients $\bar{\alpha}_j$ in D with $\nu(u) < 0$. Since $\Delta = u^{-36}\bar{\Delta}$, $0 \leq \nu(\bar{\Delta}) = \nu(\Delta) + 36\nu(u)$, i.e., $0 \leq \nu(\bar{\Delta}) < \nu(\Delta)$, therefore the normal form in the coefficient α_j and the variables w, x, y is not minimal. On the other hand, any two minimal normal forms with discriminants Δ and $\bar{\Delta}$ must have the same p-value, i.e., $\nu(\Delta) = \nu(\bar{\Delta})$. □

If D is a DEDEKIND domain of class number one with field of fractions K, for any prime p in D with valuation ν_p we have $\nu_p(\Delta) = 36\nu_p(u) + \nu_p(\bar{\Delta})$. This leads to the global definition of minimal normal form, since by means of a TSCHIRNHAUS transformation it is possible to choose a normal form with $\nu_p(\Delta)$ minimal for all primes p in D.

Definition 7.10. Let K be the field of fractions of a DEDEKIND domain D of class number one and let $\mathcal{C}$ be a PICARD curve over Spec(D). A *minimal normal form* for $\mathcal{C}$ is a normal form with all coefficients in D, such that $\nu_p(\Delta)$ is minimal among all normal forms with coefficients in D.

7.7 Good Reduction of PICARD Curves

Let $r_p(a) = a_\nu$ denote the canonical reduction modulo p defined by $D \to k_\nu = \mathcal{O}_\nu/\mathfrak{m}_\nu$, where ν is the valuation associated to the prime p, $\mathcal{O}_\nu = \{a \in K;\ \nu(a) \geq 0\}$, $\mathfrak{m}_\nu = \{a \in K;\ \nu(a) > 0\}$. For the generic fibre of a PICARD curve C_K with minimal normal form $C_K:\ f = wy^3 - (x^4 + \alpha_2 x^2 w^2 + \alpha_1 x w^3 + \alpha_0 w^4) = 0$ the reduction of C_K modulo p is given by

$$C_\nu:\ f = wy^3 - (x^4 + \alpha_{2_\nu} x^2 w^2 + \alpha_{1_\nu} x w^3 + \alpha_{0_\nu} w^4) = 0. \qquad (7.15)$$

It is a projective plane curve over k_ν. The reduction mod p is well defined up to isomorphism.

The discriminant of the reduced curve C_ν is Δ_ν, the reduction mod p is the discriminant Δ of C_K.

Definition 7.11. A PICARD curve with generic fibre over K, C_K, has *good reduction* at ν provided C_ν, the reduced curve at p, is nonsingular, otherwise we say C_K has *bad reduction* at ν.

After Section 4, if $\text{char}(k_\nu) = 3$, there exists an extension $\bar{k}_\nu$ of k_ν, such that C_ν is singular. If $\text{char}(k_\nu) \neq 3$, then C_ν is nonsingular iff $\Delta_\nu \neq 0$, hence from (7.12) C_ν has good reduction at ν iff the differences of any pair of roots of the polynomial p_4, $a_i - a_j$, $i \neq j$ are all in $\mathcal{O}_\nu^*$.

Quoting a suggestion of *F.* OORT, the finiteness theorems correspond to a generalization of an analogon to a theorem by HERMITE: fix some discrete invariants for a function field $\mathbb{G}$ over an algebraic number field K (e.g. genus, transcendence degree, etc.), impose on it some properties of good reduction corresponding to L/K being unramified in the case $[L : K] < \infty$ and show the finiteness of the number of K-isomorphism classes of such subjects (see [102], [103]).

We will prove now the conjecture of SHAFAREVIČ [103] on the finiteness of isomorphism classes of curves with good reduction outside a finite set of discrete valuations of a number field K for the cases of PICARD curves (since projective isomorphism implies K-isomorphism, it is sufficient to prove it for projective isomorphism classes).

Theorem 7.12. *Let K be a number field ($[K : \mathbb{Q}] < \infty$), T a finite set of discrete valuations of K and $Sh_{K,T}$ be the set of projective isomorphism classes of* PICARD *curves with generic fibre defined over K which have good reduction outside T. Then $\sharp Sh_{K,T} < \infty$.*

For the proof, we will need the following lemma:

Lemma 7.13. *Let T be a finite set of discrete valuations of K containing all primes dividing 2 and 3 and suppose $\mathcal{O}_{(T)}$ has class number one (it is possible adding a finite number of elements to T). Then a* PICARD *curve over K has good reduction outside T iff there exists a normal form*

$$C_K:\ f = wy^3 - (x^4 + \alpha_2 x^2 w^2 + \alpha_1 x w^3 + \alpha_0 w^4) = 0 \quad \text{with} \quad \alpha_j \in \mathcal{O}_{(T)}$$

and $\Delta \in \mathcal{O}_{(T)}^*$.

Proof: One direction of the equivalence (the sufficiency) is trivial.

Let ν be any discrete valuation of K, $\nu \notin T$, and suppose that curve C_K has good reduction at ν. Since $\nu \notin T$, ν does not divide 2 and 3, therefore the good reduction of C_K implies the existence of a normal form

$$C_K: \ f = wy^3 - (x^4 + \alpha_2 x^2 w^2 + \alpha_1 x w^3 + \alpha_0 w^4) = 0 \quad \text{with} \quad \alpha_j \in \mathcal{O}_{(T)} \qquad (7.16)$$

such that the reduced normal form

$$C_K: \ f_\nu = wy^3 - (x^4 + \alpha_{2_\nu} x^2 w^2 + \alpha_{1_\nu} x w^3 + \alpha_{0_\nu} w^4) = 0 \qquad (7.17)$$

has all its coefficients in $\mathcal{O}_\nu$ and $\Delta_\nu \in \mathcal{O}_\nu{}^*$. We can consider the coefficients of $\alpha_j \in K$ as elements of the field of quotients of $\mathcal{O}_\nu$, K_ν. Then (7.16) and (7.17) are two normal forms of a PICARD curve defined over K_ν and there exists a diagonal TSCHIRNHAUS transformation $\tau = \mathrm{diag}(1, u_\nu^3, u_\nu^4)$ relating them. So, $\alpha_{2_\nu} = u_\nu^6 \alpha_2$, $\alpha_{1_\nu} = u_\nu^9 \alpha_1$, $\alpha_{0_\nu} = u_\nu^{12} \alpha_0$, and $\Delta_\nu = u_\nu^{36} \Delta$ for some $u_\nu \in K_\nu^*$. Without loss of generality, $\alpha_{j_\nu} = \alpha_j$ for almost all $v \notin T$ (the set of primes dividing elements of T is finite), hence $u_\nu = 1$ for almost all $v \notin T$.

Since $\mathcal{O}_\nu$ has class number one, there exists $u \in K^*$, such that $\nu(u) = \nu(u_\nu)$, so we can replace above all u_ν by an unique u, for all $\nu \notin T$.

Making a TSCHIRNHAUS transformation $\tau = \mathrm{diag}(1, u^3, u^4)$ in (7.16), we obtain a normal form

$$C_K: \ \bar{f} = \bar{w}\bar{y}^3 - (\bar{x}^4 + \bar{\alpha}_2 \bar{x}^2 \bar{w}^2 + \bar{\alpha}_1 \bar{x} \bar{w}^3 + \bar{\alpha}_0 \bar{w}^4) = 0 \qquad (7.18)$$

with $u^{-6}\alpha_2 = \bar{\alpha}_2$, $u^{-9}\alpha_1 = \bar{\alpha}_1$, $u^{-12}\alpha_0 = \bar{\alpha}_0$, $u^{-36}\Delta = \bar{\Delta}$, thus $\bar{\alpha}_j \in \mathcal{O}_{(T)}$ and $\bar{\Delta} \in \mathcal{O}_{(T)}{}^*$ (since $u = u_\nu$ for all $\nu \notin T$). □

Proof of the finiteness theorem: We can assume again that $\mathcal{O}_{(T)}$ has class number one. After the lemma 7.13 above, there exists a normal form $f = wy^3 - (x^4 + \alpha_2 x^2 w^2 + \alpha_1 x w^3 + \alpha_0 w^4) = 0$ with $\alpha_j \in \mathcal{O}_{(T)}$, $\Delta \in \mathcal{O}_{(T)}{}^*$ and parameters $i, j \in \mathcal{O}_{(T)}$.

By means of a TSCHIRNHAUS transformation $\tau = \mathrm{diag}(1, u^3, u^4)$, $u \in \mathcal{O}^*_{(T)}$, the parameters i, j, Δ change over to $u^{-12} i = \bar{i}$, $u^{-18} j = \bar{j}$, $u^{-36}\Delta = \bar{\Delta}$. Therefore, the parameters i and j are determined uniquely mod $(\mathcal{O}^*_{(T)})^{12}$ and mod $(\mathcal{O}^*_{(T)})^{18}$ respectively.

Let us suppose $i, j \in \mathcal{O}^*_{(T)}$. By DIRICHLET's unit theorem for number fields [100], the quotient $\mathcal{G}(r) = \mathcal{O}^*_{(T)}/(\mathcal{O}^*_{(T)})^r$ is a finite group for $r \geq 1$. Thus, there are finite sets of representatives $\mathbb{X} = \{X_1, \ldots, X_q\}$ of $\mathcal{G}$ $(r = 12)$ and $\mathbb{Y} = \{Y_1, \ldots, Y_s\}$ of $\mathcal{G}$ $(r = 18)$, such that $i \equiv X_i \mod (\mathcal{O}^*_{(T)})^{12}$ and $j \equiv Y_j \mod (\mathcal{O}^*_{(T)})^{18}$. Since the absolute invariant i_α is equal to i^3/j^2, i_α is uniquely determined by the sets of representatives $\mathbb{X}$ and $\mathbb{Y}$ $\mathrm{mod}(\mathcal{O}^*_{(T)})^{36}$.

Hence, we can define a map φ from $Sh_{K,T}$ into a finite set

$$C_K: \ f=0 \stackrel{\varphi}{\longmapsto} \left\{(X_r, Y_s) \in \mathbb{X} \times \mathbb{Y};\ (X_r^3/Y_s^2) \equiv i_\alpha \ \mathrm{mod}(\mathcal{O}^*_{(T)})^{36}\right\}.$$

According to (7.9–7.11), $12\alpha_0 + \alpha_2^2 - 6i = 0$ and $72\alpha_0\alpha_2 - 2\alpha_2^3 - 27\alpha_1^2 - 72j = 0$; eliminating α_0 we obtain

$$\alpha_1^2 = -\frac{8}{27}(\alpha_2^3 - \frac{9}{2}i\alpha_2 + 9i)\,. \tag{7.19}$$

Thus, for any projective isomorphism class $C_K \in Sh_{K,T}$ we get a finite set of equations of type (7.19),

$$\alpha_1^2 = -\frac{8}{27}(\alpha_2^3 - \frac{9}{2}X_r\alpha_2 + 9Y_s) \tag{7.20}$$

with $(X_r, Y_r) \in \varphi(C_k)$. For $i_\alpha \not\equiv 6 \ \mathrm{mod}(\mathcal{O}^*_{(T)})^{36}$ formula (7.20) represents the equation of a (elliptic) curve of genus 1 defined over $\mathcal{O}_{(T)}$ and by SIEGEL's theorem for curves of genus 0 [Lan], (7.20) has a finite set of solutions $\alpha_1, \alpha_2 \in \mathcal{O}_{(T)}$, which determine a finite set of normal forms $f = wy^3 - (x^4 + \alpha_2x^2w^2 + \alpha_1xw^3 + \alpha_0w^4) = 0$ with $\alpha_j \in \mathcal{O}^*_{(T)}$, since $\alpha_0 = (6X_i - \alpha_2^2)/12$.

The case $i_\alpha \equiv 6 \ \mathrm{mod}(\mathcal{O}^*_{(T)})^{36}$ can be excluded if $\mathcal{O}_{(T)}$ is infinite (for $\mathcal{O}_{(T)}$ finite there is no need to prove a finiteness theorem). In fact, if $i_\alpha \equiv 6 \ \mathrm{mod}(\mathcal{O}^*_{(T)})^{36}$, say $i_\alpha = 6u^{36}$ with $u \in \mathcal{O}_{(T)}$, then $\Delta, j, 6 \in \mathcal{O}_{(T)}$ implies $(u^{36} - 1) \in \mathcal{O}^*_{(T)}$, since $\Delta = 6j^2(u^{36} - 1)/27$. Thus, $u^{36} \in J_{K,T} = \{\lambda; \lambda \in \mathcal{O}^*_{(T)}$ and $(\lambda - 1) \in \mathcal{O}^*_{(T)}\}$. The set $J_{K,T}$ is finite (see [100], VII.4), therefore the number of elements belonging to the equivalence class of $1 \ \mathrm{mod}(\mathcal{O}^*_{(T)})^{36}$ is finite ($i, j \in \mathcal{O}_{(T)}$ implies $6/i_\alpha \equiv 1 \ \mathrm{mod}(\mathcal{O}^*_{(T)})^{36}$) and from the finiteness of the number of equivalence classes in $\mathcal{G}$ ($r = 36$) follows the finiteness of $Sh_{K,T}$.

If $i \notin \mathcal{O}^*_{(T)}$ or $j \notin \mathcal{O}^*_{(T)}$ but $i, j \neq 0$, consider a common denominator u of i and j (i.e., $ui, uj \in \mathcal{O}^*_{(T)}$) and the finite set $T' = T \cup \{\text{all primes dividing } u\}$. Then, we can suppose that $\mathcal{O}_{(T')}$ has also class number one and we obtain, after a diagonal TSCHIRNHAUS transformation $\tau = \mathrm{diag}(1, u^3, u^4)$, a normal form $f = wy^3 - (x^4 + \alpha_2x^2w^2 + \alpha_1xw^3 + \alpha_0w^4) = 0$ with $\alpha_j \in \mathcal{O}_{(T')}$ and $i, j, \Delta \in \mathcal{O}^*_{(T')}$ and the result obtained above also holds.

$i = 0$ and $j = 0$ can not hold simultaneously ($\Delta \neq 0$), thus $i = 0$ (or $j = 0$, respectively) implies that (7.19) is a curve of genus 1 defined over $\mathcal{O}_{(T)}$ and has a finite set of solutions (α_1, α_2), which determine a finite set of coefficients $(\alpha_0, \alpha_1, \alpha_2)$ for the normal forms of the projective isomorphism classes of $Sh_{K,T}$.

By means of BAKER's method (see [Bak], [Bri] and [Cas]) the solutions of equation (7.20) can be effectively obtained, in case it represents a curve of genus > 0. □

J. Estrada-Sarlabous, ICIMAF. Academy of Sciences of Cuba,
Calle E No.309, esquina a 15, Vedado. La Habana 4. Cuba.

8 Appendix* II
The Hilbert Problems 7, 12, 21 and 22

8.1 Irrationality and Transcendence of Certain Numbers

HERMITE's arithmetical theorems on the exponential function and their extension by LINDEMANN are certain of the admiration of all generations of mathematicians. Thus the task at once presents itself to penetrate further along the path here entered, as A. HURWITZ has already done in two interesting papers, "Über arithmetische Eigenschaften gewisser transzendenter Funktionen". I should like, therefore, to sketch a class of problems which, in my opinion, should be attacked as here next in order. That certain special transcendental functions, important in analysis, take algebraic values for certain algebraic arguments, seems to us particularly remarkable and worthy of thorough investigation. Indeed, we expect transcendental functions to assume, in general, transcendental values for even algebraic arguments; and, although it is well known that there exist integral transcendental functions which even have rational values for all algebraic arguments, we shall still consider it highly probable that the exponential function $e^{i\pi x}$, for example, which evidently has algebraic values for all rational arguments z, will on the other hand always take transcendental values for irrational algebraic values of the argument z. We can also give this statement a geometrical form, as follows: *If, in an isosceles triangle, the ratio of the base angle to the angle at the vertex be algebraic but not rational, the ratio between base and side is always transcendental.*

In spite of the simplicity of this statement and of its similarity to the problems solved by HERMITE and LINDEMANN, I consider the proof of this theorem very difficult; as also the proof that *the expression* a^β, *for an algebraic base* a *and an irrational algebraic exponent* β, *e.g., the number* $2^{\sqrt{s}}$ *or* $e^\pi = i^{-2i}$, *always represents a transcendental or at least an irrational number.* It is certain that the solution of these and similar problems must lead us to entirely new methods and to a new insight into the nature of special irrational and transcendental numbers.

8.2 Extension of KRONECKER's Theorem on Abelian Fields to any Algebraic Realm of Rationality.

The theorem that every abelian number field arises from the realm of rational numbers by the composition of fields of roots of unity is due to KRONECKER. This fundamental theorem in the theory of integral equations contains two statements, namely:

---

*) Translation taken from [15]

First. It answers the question as to the number and existence of those equations which have a given degree, a given abelian group and a given discriminant with respect to the realm of rational numbers.

Second. It states that the roots of such equations form a realm of algebraic numbers which coincides with the realm obtained by assigning to the argument z in the exponential function $e^{i\pi z}$ all rational numerical values in succession.

The first statement is concerned with the question of the determination of certain algebraic numbers by their groups and their branching. This question corresponds, therefore, to the known problem of the determination of algebraic functions corresponding to given RIEMANN surfaces. The second statement furnishes the required numbers by transcendental means, namely, by the exponential function $e^{i\pi z}$.

Since the realm of the imaginary quadratic number fields is the simplest after the realm of rational numbers, the problem arises, to extend KRONECKER's theorem to this case. KRONECKER himself has made the assertion that the abelian equations in the realm of a quadratic field are given by the equations of transformation of elliptic functions with singular moduli, so that the elliptic function assumes here the same role as the exponential function in the former case. The proof of KRONECKER's conjecture has not yet been furnished; but I believe that it must be obtainable without very great difficulty on the basis of the theory of complex multiplication developed by H. WEBER with the help of the purely arithmetical theorems on class fields which I have established.

Finally, the extension of KRONECKER's theorem to the case that, *in place of the realm of rational numbers or of the imaginary quadratic field, any algebraic field whatever is laid down as realm of rationality*, seems to me of the greatest importance. I regard this problem as one of the most profound and far reaching in the theory of numbers and of functions.

The problem is found to be accessible from many stand-points. I regard as the most important key to the arithmetical part of this problem the general law of reciprocity for residues of l-th powers within any given number field.

As to the function-theoretical part of the problem, the investigator in this attractive region will be guided by the remarkable analogies which are noticeable between the theory of algebraic functions of one variable and the theory of algebraic numbers. HENSEL has proposed and investigated the analogue in the theory of algebraic numbers to the development in power series of an algebraic function; and LANDSBERG has treated the analogue of the RIEMANN-ROCH theorem. The analogy between the deficiency of a RIEMANN surface and that of the class number of a field of numbers is also evident. Consider a RIEMANN surface of deficiency $p = 1$ (to touch on the simplest case only) and on the other hand a number field of class $h = 2$. To the proof of the existence of an integral everywhere finite on the RIEMANN surface, corresponds the proof of the existence on an integer a in the number field such that the number $\sqrt{a}$ represents a quadratic field, relatively unbranched with respect to the fundamental field. In the theory of algebraic functions, the method of boundary values (*Randwertaufgabe*) serves, as is well known, for the proof of RIEMANN's existence theorem. In the theory of number fields also, the proof of the existence of just this number a offers the greatest difficulty. This proof succeeds with indispensable assistance from the theorem that in the number field there are always prime ideals corresponding to given residual properties. This latter fact is therefore the analogue in number theory to the problem of boundary values.

The equation of ABEL's theorem in the theory of algebraic functions expresses, as is well known, the necessary and sufficient condition that the points in question on the RIEMANN surface are the zero points of an algebraic function belonging to the surface. The exact analogue of ABEL's theorem, in the theory of the number field of class $h = 2$, is the equation of the law of quadratic reciprocity $\left(\frac{a}{j}\right) = +1$, which declares that the ideal j is then and only then a principal ideal of the number field when the quadratic residue of the number a with respect to the ideal j is positive.

It will be seen that in the problem just sketched the three fundamental branches of mathematics, number theory, algebra and function theory, come into closest touch with one another, and *I am certain that the theory of analytical functions of several variables in particular would be notably enriched if one should succeed in finding and discussing those functions which play the part for any algebraic number field corresponding to that of the exponential function in the field of rational numbers and of the elliptic modular functions in the imaginary quadratic number field.*

8.3 Proof of the Existence of Linear Differential Equations Having a Prescribed Monodromic Group

In the theory of linear differential equations with one independent variable z, I wish to indicate an important problem, one which very likely RIEMANN himself may have had in mind. This problem is as follows: *To show that there always exists a linear differential equation of the Fuchsian class, with given singular points and monodromic group.* The problem requires the production of n functions of the variable z, regular throughout the complex z plane except at the given singular points; at these points the functions may become infinite of only finite order, and when z describes circuits about these points the functions shall undergo the prescribed linear substitutions. The existence of such differential equations has been shown to be probable by counting the constants, but the rigorous proof has been obtained up to this time only in the particular case where the fundamental equations of the given substitutions have roots all of absolute magnitude unity. L. SCHLESINGER has given this proof, based upon POINCARÉ's theory of the Fuchsian ζ-functions. The theory of linear differential equations would evidently have a more finished appearance if the problem here sketched could be disposed of by some perfectly general method.

8.4 Uniformization of Analytic Relations by Means of Automorphic Functions

As POINCARÉ was the first to prove, it is always possible to reduce any algebraic relation between two variables to uniformity by the use of automorphic functions of one variable. That is, if any algebraic equation in two variables be given, there can always be found for these variables two such single valued automorphic functions of a single variable that their substitution renders the given algebraic equation an identity. The generalization of this fundamental theorem to any analytic non-algebraic relations whatever between two variables has likewise been attempted with success by POINCARÉ, though by a way entirely different from that which served him in the special problem first mentioned. From POINCARÉ's proof of the possibility of reducing to uniformity an arbitrary analytic relation between two variables, however, it does not become apparent whether the resolving functions can be determined to meet certain additional conditions. Namely, it is not shown whether the two single valued functions of the one new variable can be chosen that, while this variable traverses the *regular* domain of those functions, the totality of all regular points of the given analytic field are actually reached and represented. On the contrary it seems to be the case, from POINCARÉ's investigations, that there are beside the branch points certain others, in general infinitely many other discrete exceptional points of the analytic field, that can be reached only by making the new variable approach certain limiting points of the functions. In view of the fundamental importance of POINCARÉ's formulation of the question it seems to me that an elucidation and resolution of this difficulty is extremely desirable.

In conjunction with this problems comes up *the problem of reducing to uniformity an algebraic or any other analytic relation among three or more complex variables — a problem which is known to be solvable in many particular cases. Toward the solution of this the recent investigations of* PICARD *on algebraic functions of two variables are to be regarded as welcome and important preliminary studies.*

Basic Notations

- $\mathbb{C}, \mathbb{R}, \mathbb{Q}, \bar{\mathbb{Q}}, \mathbb{Z}$: complex, real, algebraic numbers, integers;
- $\mathbb{P}^N$: (complex) projective space of dimension N;
- $\mathbb{P}^N(L)$: points of $\mathbb{P}^N$ with coordinates in the field $L \subseteq \mathbb{C}$;
- $V(L) = V \cap \mathbb{P}^N(L)$, V a subvariety of $\mathbb{P}^N$;
- $\mathbb{G}l_n(\mathbb{G}l_n(R))$: general linear group (with coefficients in R, R a subring of $\mathbb{C}$);
- $\mathbb{PG}l_n$: the projective group $\mathbb{G}l_n/\mathbb{G}l_1$;
- $\mathbb{P}G$: the image of G in $\mathbb{PG}l_n$, G a subgroup of $\mathbb{G}l_n$;
- $\mathbb{U}((2,1), A)$: the unitary group of a hermitian form of signature (2,1) with coefficients in the ring $A \subseteq \mathbb{C}$ closed under complex conjugation;
- $\mathcal{O}_L$: ring of integers in the number field L,
- $\Gamma_L = \mathbb{U}((2,1), \mathcal{O}_L)$ the full PICARD modular group of the imaginary quadratic number field L;
- $K = \mathbb{Q}(\sqrt{-3}) = \mathbb{Q}(\rho)$: the field of EISENSTEIN numbers, ρ the primitive third unit root $e^{2\pi i/3}$;
- $\mathbb{B}^2 = \{(z_1, z_2) \in \mathbb{C}^2; |z_1|^2 + |z_2|^2 < 1\}$ the standard complex 2-ball;
- $j(\tau) = q^{-1} + 744 + 196884q + 21493760q^2 + \ldots$, $q = \exp(2\pi i\tau)$ the elliptic modular function defined on:
- $\mathbb{H} = \{\tau \in \mathbb{C}; Im\,\tau > 0\}$ the POINCARÉ upper half plane;
- $\mathbb{H}_g = \{Z \in \mathbb{G}l_g(\mathbb{C});\ Z = {}^tZ,\ Im\,Z > 0\}$ the SIEGEL upper half space;
- $\mathbb{S}p(2g, A)$: symplectic group acting on $\mathbb{H}_g$, $A = \mathbb{R}$ or $\mathbb{Z}$;
- $\mathcal{A}_g = \mathbb{S}p(2g, \mathbb{Z})\backslash\mathbb{H}_g$: the (non-compact) moduli space of g-dimensional (principally) polarized abelian varieties.
- ——> morphism
- ---> map (rational or multivalued)

Bibliography

[1] Alezais, M.R., *Sur une classe de fonctions hyperfuchsiennes*, Ann. Ec. Norm. **19**, Sér. 3 (1902), 261–323.

[2] Baily, W.L., Borel, A., *Compactification of arithmetic quotients of bounded symmetric domains*, Ann. of Maths. **84** (1966), 442–528.

[3] Baker, A. (Ed.), *New advances in transcendence theory*, Cambridge Univ. Press, Cambridge 1988.

[4] Baker, A., Wuestholz, G., *Logarithmic Forms*, Preprint, ETH Zürich 1990.

[5] Barthel, G., Hirzebruch, F., Hoefer, Th., *Geradenkonfigurationen und algebraische Flächen*, Asp. Math. **D4**, Vieweg, Braunschweig 1987.

[6] Beukers, F., *Some new results on algebraic independence of E-functions*, in [3], 56–67.

[7] Beukers, F., Wolfart, J., *Algebraic values of hypergeometric functions*, in [3], 68–81.

[8] Birch, B.J., Kuyck, W. (Ed.), *Modular functions of one variable IV*, Lecture Notes in Math. **476**, Springer 1975.

[9] Borel, A., *Compact Clifford-Klein forms of symmetric spaces*, Topology **2** (1963), 111–122.

[10] Borel, A., *Some metric properties of arithmetic quotients of symmetric spaces and an extension theorem*, J. Diff. Geom **6** (1972), 543–560.

[11] Borevič, S.I., Shafarevič, I.R., *Number theory*, Nauka, Moscow 1985.

[12] Bost, J.B., *Fonctions de Green-Arakelov, fonctions theta et courbes de genre 2*, C.R. Acad. Sci., Paris, Ser. 1 **305** (1987), 643–646.

[13] Bost, J.B., Mestre, J.F., Moret-Bailly, L., *Sur le calcul explicite des "classes de Chern" des surfaces arithmétiques de genre 2*, in [86], 69–105.

[14] Brieskorn, E., *Rationale Singularitäten komplexer Flächen*, Inv. math. **4** (1967), 336–358.

[15] Browder, F.E. (Ed.), *Mathematical developments arising from Hilbert problems*, Proc. Symp. Pure Math. **28**, parts, 1, 2, AMS, Providence 1976.

[16] Cassels, J.W.S., Froehlich, A. (Ed.), *Algebraic number theory*, Academic Press, London – New York 1967.

[17] Cornell, G., Silverman, J.H. (Ed.), *Arithmetic Geometry*, Springer, New York 1986.

[18] DELIGNE, P., *Equations différentielles à points singuliers réguliers*, SLN **163**, Springer 1970.

[19] DELIGNE, P., *Travaux de Griffiths*, Sém. Bourbaki May 1970, Exp. **376**, SLN **180**, Springer 1971.

[20] DELIGNE, P., *Travaux de Shimura*, Séminaire Bourbaki 1970/71, Exp. **389**, SLN **244**. Springer 1971.

[21] FALTINGS, G., *Calculus on arithmetic surfaces*, Ann. Math. **119** (1984), 387–424.

[22] FALTINGS, G., *Endlichkeitssätze für abelsche Varietäten über Zahlenkörpern*, Inv. Math. **73** (1983), 349–366.

[23] FEUSTEL, J.-M., *Ringe automorpher Formen auf der komplexen Einheitskugel und ihre Erzeugung durch Thetakonstanten*, Prepr. Ser. P-Math-**13**, AdW DDR, IMath (1986).

[24] GELFOND, A.O., *Sur le septième problème de Hilbert*, Izv. Akad. Nauk SSSR, ser. Phys.-Math. **4** (1934), 623–630.

[25] GRIFFITHS, PH.A., *Complex analytic properties of certain Zariski open sets on algebraic varieties*, Ann. Math. **94** (1971), 21–51.

[26] GRIFFITHS, PH.A., HARRIS, J., *Principles of algebraic geometry*, Wiley, New York – Chichester – Brisbane – Toronto 1978.

[27] HARTSHORNE, F., *Algebraic Geometry*, Grad. Texts in Math. **52**, Springer, Berlin – Heidelberg – New York 1977.

[28] HASSE, H., *History of class field theory, Ch. XI in [16].*

[29] HERMITE, CH., *Sur la fonction exponentielle*, C.R. Acad. Sci. **77** (1873), Paris, 18–24, 74–79, 226–233, 285–293; Œuvres III, 150–181.

[30] HILBERT, D., *Mathematische Probleme*, Nachr. Kgl. Ges. d. Wiss. Goettingen, Math.-Phys. Kl. **3** (1900), 253–297.

[31] HILBERT, D., *Die Hilbertschen Probleme*, Ostwalds Klassiker der exakten Wissenschaften, Geest & Portig, Leipzig 1971.

[32] HOLZAPFEL, R.-P., *Ball cusp singularities*, Nova acta Leopoldina, NF 52, Nr. **240** (1981), 109–117.

[33] HOLZAPFEL, R.-P., *Geometry and arithmetic around Euler partial differential equations*, Dt. Verl. d. Wiss., Berlin / Reidel, Dordrecht 1986.

[34] HOLZAPFEL, R.-P., *An arithmetic uniformization for arithmetic points of the plane by singular moduli*, Journ. Ramanujan Math. Soc. **3**, No. 1 (1988), 35–62.

[35] HOLZAPFEL, R.-P., *On the Nebentypus of Picard Modular Forms*, Math. Nachr. **139** (1988), 115–137.

[36] HOLZAPFEL, R.-P., *Discrete analysis of surface coverings I*, Rev. Roum. Math. **33** (3) (1988), 197–232.

[37] HOLZAPFEL, R.-P., *Discrete analysis of surface coverings II*, Rev. Roum. Math. **33** (4) (1988), 305–348.

[38] HOLZAPFEL, R.-P., *Volumes of fundamental domains of Picard modular groups*, Proc. Workshop Arithm. of Complex manifolds, SLN on Math. **1399**, Springer 1988.

[39] HOLZAPFEL, R.-P., *An effective finiteness theorem for ball lattices*, Schriftenr. Univ. Erlangen Nr. 41 (Forsch. Schwp. Komplexe Mannigf.), 1989, SLN **1447**, Springer 1990, 203–236.

[40] HUNT, B., *A Siegel modular threefold that is a Picard modular threefold*, Comp. Math. **76** (1990), 203–242.

[41] HUSEMOLLER, D., *Elliptic curves*, Grad. Texts in Math. **111**, Springer, Berlin – Heidelberg – New York 1986.

[42] KATZ, N.M., *Algebraic solutions of differential equations (p-curvature and the Hodge filtration)*, Inv. math. **18** (1972), 1–118.

[43] KOBAYASHI, R., *Uniformization of complex surfaces*, Preprint, Univ. Tokyo 1990.

[44] LANG, S., *Transcendental points on group varieties*, Topology **1** (1962), 313–318.

[45] LANG, S., *Introduction to algebraic and abelian functions*, Addison-Wesley, Reading 1972.

[46] LANG, S., *Elliptic functions*, Addison Wesley, London – Amsterdam – Tokyo 1973.

[47] LANG, S., *Fundamentals of diophantine geometry*, Springer, New York 1983.

[48] LANG, S., *Complex multiplication*, Grundl. Math. Wiss. **255**, Springer 1983.

[49] LANG, S., *Introduction to Arakelov theory*, Springer, New York 1988.

[50] LINDEMANNN, F., *Ueber die Zahl* π, Math. Ann. **20** (1882), 213–225.

[51] MANIN, YU.I., *Algebraic curves over fields with differentiation*, Izv. Acad. Nauk SSSR, Ser. Mat. **22** (1958), 737–756.

[52] MILNE, J.S., *Points on Shimura varieties mod* p, Proc. Symp. Pure Math. **33** (1979), vol. 2, 165–184.

[53] MILNE, J.S., *Jacobian Varieties*, Ch. VII in [17].

[54] MORET-BAILLY, L., *La formule de Noether pour les surfaces arithmétiques*, Inv. Math. **98** (1989), 491–498.

[55] MUMFORD, D., *Abelian varieties*, Tata Lectures, Bombay 1968.

[56] MUMFORD, D., *Tata Lectures on Theta I*, Birkhäuser, Boston – Basel – Stuttgart 1983.

[57] NEUKIRCH, J., *Class field theory*, Grundl. d. math. Wiss. **280**, Springer 1986.

[58] OGATA, S., *Signature defects and eta invariants of Picard modular cusp singularities*, J. Math. Soc. Japan **42** (1990), 659–675.

[59] PARSHIN, A.N., *On the use of branched coverings in the theory of diophantine equations*, Mat. Sbornik **180** (1989), 244–259.

[60] PICARD, E., *Sur des fonctions de deux variables indépendantes analogues aux fonctions modulaires*, Acta math. **2** (1883), 114–135.

[61] PICARD, E., *Sur les formes quadratiques ternaires indéfinies et sur les fonctions hyperfuchsiennes*, Acta math. **5** (1884), 121–182.

[62] REIMANN, H., *Das kanonische Modell eines Kugelquotienten*, Math. Nachr. (1991).

[63] REINER, I., *Maximal orders*, Academic Press, London – New York – San Francisco 1975.

[64] ROEHRL, H., *Das Riemann-Hilbertsche Problem der Theorie der linearen Differentialgleichungen*, Math. Ann. **133** (1957), 1–25.

[65] SCHNEIDER, T., *Transzendenzuntersuchungen periodischer Funktionen I. Transzendenz von Potenzen*, Journ. f. reine u. angew. Math. **172** (1934), 65–69.

[66] SCHNEIDER, T., *Arithmetische Untersuchungen elliptischer Integrale*, Math. Ann. **113** (137), 1–13.

[67] SCHNEIDER, T., *Zur Theorie der Abelschen Funktionen und Integrale*, J. reine angew. Math. **183** (1941), 110–128.

[68] SCHNEIDER, T., *Einführung in die transzendenten Zahlen*, Grdl. d. Math. Wiss. **81**, Berlin – Goettingen – Heidelberg 1957.

[69] SERRE, J.P., *Abelian ℓ-adic representations and elliptic curves*, Benjamin, New York – Amsterdam 1968.

[70] SERRE, J.P., *Quelques Applications du théorème de densité de Chebotarev*, Publ. Math. I.H.E.S., **54** (1981), 123–201.

[71] SERRE, J.P., *Lectures on the Mordell-Weil Theorem*, Asp. of Math. E15, Vieweg, Braunschweig 1989.

[72] SHIDLOVSKY, A.B., *Transcendental numbers*, Nauka, Moscow 1987, de Gruyter, Berlin – New York 1989.

[73] SHIGA, H., *On the construction of algebraic numbers as special values of the Picard modular function*, Preprint, Chiba Univ. 1987.

[74] SHIGA, H., *On the representation of the Picard modular function by theta constants* I–II, Publ. RIMS, Kyoto Unvi. **24** (1988), 311–360.

[75] SHIGA, H., *Transcendency for the values of certain modular functions of several variables at algebraic points*, Preprint, Chiba Univ. 1990.

[76] SHIGA, H., *On the transcendency of the values of the modular function at algebraic points*, Preprint, Chiba Univ. 1991.

[77] SHIGA, H., *On the transcendency of the modular function at algebraic points*, in [92], 103–107.

[78] SHIMURA, G., *Introduction to the arithmetic theory of automorphic functions*, Iwanami Shoten - Princeton UP 1971.

[79] SHIMURA, G., TANIYAMA, Y., *Complex multiplication of abelian varieties and its applications to number theory*, Publ. Math. Soc. Japan **6**, 1961.

[80] SIEGEL, C.-L., *Über einige Anwendungen diophantischer Approximationen*, Abh. Preuss. Akad. Wiss., Phys. Math. **1** (1929), 1–70.

[81] SIEGEL, C.-L., *Über die Perioden elliptischer Funktionen*, J. f. reine angew. Math. **167** (1932), 62–69.

[82] SIEGEL, C.-L., *Transcendental numbers*, Ann. of Math. Studies **16**, Princeton UP 1949.

[83] SILVERMAN, J.H., *The classical theory of heights*, ch. VI in [17].

[84] SILVERMAN, J.H., *Heights and elliptic curves*, ch. X in [17].

[85] SILVERMAN, J.H., *A survey of the theory of heights*, Contemp. Math. **67** (1987), 269–278.

[86] SZPIRO, L. (Ed.), *Les pinceaux de courbes elliptiques*, Sém. Paris 1988, Astérisque **183**, 1990.

[87] TAKAGI. T., *Über eine Theorie des relativ-Abelschen Zahlkörpers*, Journ. Coll. Science **41** (1920), 1–132.

[88] TATE, J., *Algorithm for finding the type of a singular fibre in an elliptic pencil*, in [8], 33–52.

[89] VOJTA, P., *Diophantine inequalities and Arakelov theory*, Appendix in LANG's book [49].

[90] WÜSTHOLZ, G., *Algebraische Punkte auf analytischen Untergruppen algebraischer Gruppen*, Ann. of Math. **129** (1989), 501–517.

[91] YOSHIDA, M., *Fuchsian Differential Equations*, Aspects of Math., Vieweg, Braunschweig 1987.

[92] YOSHIDA, M., (Ed.), *Special Differential Equations*, Proc. of the Taniguchi Workshop, Katata, Dep. Math. Kyushu Univ. Fukuoka 1991.

Additional References for Appendix I

[Bak] BAKER, A., *Transcendental number theory*, Cambridge Univ. Press, London-New York 1975.

[Bri] BRINDZA, B., *On S-integral solutions of the equation* $y^m = f(x)$, Acta math.Hung. 44/1–2 (1984), 1–12.

[Cas] CASSELS, J.W.S., *Explicit results on the arithmetic of curves of higher genus*, Russian Math. Surveys 40:4 (1985), 43–48

[Cle] CLEBSCH, A., *Theorie der binären quadratischen Formen*, Leipzig, 1872.

[Dix] DIXMIER, J., *On the projective invariants of quadratic plane curves*, Advances in Math. 64, (1987), 279–304

[Est] ESTRADA-SARLABOUS, J., *Higher differentials on cyclic curves*, Math. Nachrichten 135, (1988), 311–317

[FW] FALTINGS, G., WÜSTHOLZ, G., ET AL., *Rational Points*, Aspects of Math. E6 (1986), Vieweg Braunschweig-Wiesbaden

[Lan] LANG S., *Diophantine Geometry*, Interscience Publ. (1962)

[MF] MUMFORD, D., FOGARTY, J., *Geometric Invariant Theory*, 2-nd edition, Springer-Verlag, 1982

[Oor] OORT, F., *Hyperelliptic Curves over Number Fields*, Springer Lecture Notes **412**, 1972.

[Sha] SHAFAREVIČ, I.R., Proc. ICM, Stockholm 1962, 163–176 (AMS Translations 31 (1963), 25–39)

Index